Praise for *Mixed Signals*

"With this work the torch passes to a new generation of SETI historians, who analyze not only the science of the search for extraterrestrial intelligence but also its cultural, religious, and political aspects. Focusing on the 1950s through the 1980s, Rebecca Charbonneau brilliantly explores both human and extraterrestrial communication, while vividly portraying CETI/SETI in the context of the Cold War."

Steven J. Dick, former NASA Chief Historian and
author of *Astrobiology, Discovery, and Societal Impact*

"Charbonneau has accomplished here the rare hat-trick of innovative research, incisive argument, and delightful writing, making this book an invigorating pleasure to read and a vital view of science history to engage with. A must-read for scientists, historians, and anyone curious about what – and whom – we seek in the stars. Charbonneau offers a new and compelling way to understand the search."

Jaime Green, author of *The Possibility of Life:
Science, Imagination, and Our Quest for Kinship in the Cosmos*

"Does intelligent life exist beyond our planet? Scientists have been searching for evidence of it for decades. Now historian and SETI researcher Rebecca Charbonneau offers an engrossing and surprising history of those efforts on both sides of the Iron Curtain and shows that what we seek in outer space has repeatedly led us right back to Earth."

Greg Eghigian, author of *After the Flying Saucers Came:
A Global History of the UFO Phenomenon*

"Any scientist who attempts to find, or communicate with, alien life possesses curiosity about and openness to hypothetical lives lived very differently from their own. In *Mixed Signals*, Rebecca Charbonneau offers the first investigation into those scientific attempts as they played out during the Cold War, among American and Soviet humans who were often alien to each other. *Mixed Signals* is an insightful, rigorously researched history that swirls the celestial and the terrestrial together. Too often, astronomical science is divorced from the earthly conditions it inhabits, eschewing politics for purity; *Mixed Signals* is an antidote to that attitude, showing science's influence on politics, politics' influence on science, and the overlap between communicating on Earth and communicating to the cosmos. The events and people detailed in this book show that humans' attempts to learn about life in the universe are, in the end, fundamentally about our home planet."

Sarah Scoles, science journalist and author of *Making Contact:*
Jill Tarter and the Search for Extraterrestrial Intelligence

"With clarity and insight, Rebecca Charbonneau has given us a beautifully written tour of the early history of SETI. More than just a description of who did what when, Charbonneau reveals the intricate webs of influence that knitted the deepest questions scientists can ask (Are we alone?) together with the prosaic realities of international politics and conflict. A must-read for anyone interested in SETI or in the intersections of science and culture."

Adam Frank, University of Rochester and
author of *The Little Book of Aliens*

"Charbonneau offers a truly fresh take on the often-told story of the origins of SETI, the Search for Extraterrestrial Intelligence. Through interviews and examination of the historical record, she has unearthed fascinating anecdotes about Sagan, Drake, Dyson, Shklovskii, and other SETI pioneers that illuminate how the field emerged from the military and political tangles of the Cold War. Charbonneau shows how the problem of communication and collaboration across the Iron Curtain closely mirrored the problem of communication with alien life. An instant must-read for students of SETI and those who want to know how the quest to answer the biggest question in astronomy truly began."

Jason T. Wright, Director of the Penn State
Extraterrestrial Intelligence Center

"A fascinating dive into the Cold War-era history of humanity's search for alien life, filled with a treasure trove of remarkable events and encounters and supported by meticulous research. Charbonneau brings to life the personalities who gave rise to the modern age of SETI. She highlights how SETI is as much a search for the human condition as it is for alien life."

David Kipping, Columbia University

Mixed Signals

This book is dedicated to Ken Kellermann

MIXED SIGNALS

Alien Communication Across the Iron Curtain

Rebecca Charbonneau

polity

Copyright © Rebecca Charbonneau 2025

The right of Rebecca Charbonneau to be identified as Author of this Work has been asserted in accordance with the UK Copyright, Designs and Patents Act 1988.

First published in 2025 by Polity Press

Polity Press
65 Bridge Street
Cambridge CB2 1UR, UK

Polity Press
111 River Street
Hoboken, NJ 07030, USA

All rights reserved. Except for the quotation of short passages for the purpose of criticism and review, no part of this publication may be reproduced, stored in a retrieval system or transmitted, in any form or by any means, electronic, mechanical, photocopying, recording or otherwise, without the prior permission of the publisher.

ISBN-13: 978-1-5095-5691-5

A catalogue record for this book is available from the British Library.

Library of Congress Control Number: 2024937008

Typeset in 11.5 on 14pt Adobe Garamond
by Cheshire Typesetting Ltd, Cuddington, Cheshire
Printed and bound in Great Britain by CPI Group (UK) Ltd, Croydon

The publisher has used its best endeavours to ensure that the URLs for external websites referred to in this book are correct and active at the time of going to press. However, the publisher has no responsibility for the websites and can make no guarantee that a site will remain live or that the content is or will remain appropriate.

Every effort has been made to trace all copyright holders, but if any have been overlooked the publisher will be pleased to include any necessary credits in any subsequent reprint or edition.

For further information on Polity, visit our website:
politybooks.com

Contents

A Note on Terminology	viii
Foreword	xi
Introduction	1
1. What Is CETI? Reframing the History of Communication with Extraterrestrial Intelligence	13
2. Alien Intelligence: Radio Astronomy and the Dawn of the Cold War	43
3. Telegram Killed the Radio Star: The First "False Alarm" in CETI	76
4. First Contact: The Relationship between Carl Sagan and I. S. Shklovsky	102
5. Why Can't We Be Friends? Messaging Extraterrestrial and Terrestrial Intelligence	131
6. The Cosmic Prism: CETI and Existential Threat	162
Conclusion	190
Notes	205
Index	233

A Note on Terminology

This book covers the history of the scientific effort to search for and communicate with extraterrestrial intelligence during the Cold War period. Over the past 60+ years, the terms used to describe this field of science have evolved. I will therefore use three different sets of terms in this book. The rationale for these distinctions will be explained in greater detail in Chapter 1.

When describing events, people, and themes that pertain to the years between 1959 and 1975, the acronym CETI (communication with extraterrestrial intelligence) will be used. CETI is the major topic of concern in this book.

When describing events, people, and themes that pertain to the 1980s and onwards, the acronym SETI (search for extraterrestrial intelligence) will be used.

When describing events, people, and themes that encompass the entire field, spanning 60+ years, the acronym C/SETI will be used.

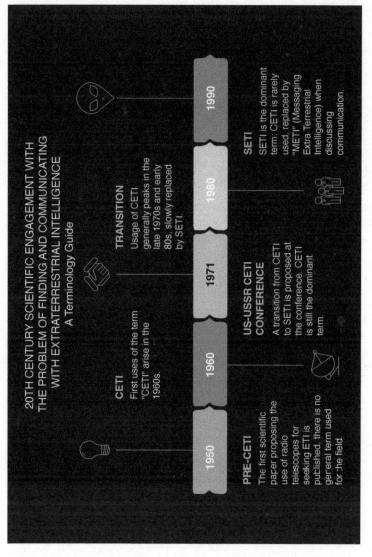

A Note on Terminology

Foreword

This is a book that, on the surface, concerns the history of the search for and communication with extraterrestrial intelligence, also known respectively as SETI and CETI. As you will soon learn, CETI developed in the mid-twentieth century, out of a field of astronomy known as radio astronomy. Yet today many radio astronomers do not professionally engage in CETI, largely considering it a fringe subject – maybe even a pseudoscience, as it is to some.

To illustrate this point, I took a few hours to run around my workplace, the National Radio Astronomy Observatory (NRAO) headquarters in Charlottesville, Virginia. I asked my astronomy colleagues this simple question: "What are your thoughts on CETI/SETI?" It was not a comprehensive sociological survey by any means, of course, but here are a few anecdotal replies that I found generally representative of the answers I received that day:

> It is an important enough question that it's worth devoting some resources to it. But it's unlikely we will ever see a positive result in our lifetimes. I think we are probably alone in the universe. (Radio astronomer)[a]

> Do I think there is intelligent life in the universe? Absolutely. Will we be able to make contact? Probably not. (Data analyst for the ALMA radio telescope)

> What's SETI? (Astrochemistry graduate student)

So why write a book about a subject that seemingly does not matter much to the professional astronomy community? I might begin by

---

[a] This radio astronomer also happens to be my husband, who clearly made a mistake in marrying an optimistic SETI researcher!

noting that there was a time when nobody was taking radio astronomy very seriously either. In fact, even NRAO's first director, the highly esteemed optical astronomer Otto Struve, had little faith in the new science when he took up his post in 1959. NRAO radio astronomer and CETI scientist Frank Drake later recalled his interactions with Struve in the early 1960s:

> He thought he could predict the future of radio astronomy, and it looked bleak to him. He warned us that unless some exciting discovery turned up – something like a variable radio source ... we would run out of things to observe in one year and then we'd all be out of work.[1]

But our preconceived beliefs about what is "out there" do not always line up with the reality of what we find. Despite Struve's concerns, the radio universe turned out to be an exciting place (and we all still have our jobs, seven decades later). And CETI is part of the reason Struve eventually became excited about his job at NRAO.

The first scientific search for extraterrestrial intelligence was conducted at NRAO in 1960 (you will learn more about this project shortly) and it energized Struve. The project, called Project Ozma, used the Observatory's 85-1 radio telescope to search for intelligent extraterrestrial signals from two nearby sun-like stars. Struve published an article in the magazine *Physics Today* shortly after this search and remarked on the surprisingly polarizing reactions to the new field of CETI, noting:

> This project has been given an unreasonable amount of publicity, often incorrect or distorted and always with the wrong emphasis. It has aroused more vitriolic criticisms and more laudatory comments than any other recent astronomical venture, and it has divided the astronomers into two camps: those who are all for it and those who regard it as the worst evil of our generation.[2]

A divisive science for a divisive time; the tension between two diametrically opposed viewpoints described by Struve parallels the Cold War environment that both CETI and radio astronomy grew up in. This book hopes to highlight how the Cold War shaped the development of CETI and radio astronomy. To this end, it will focus especially on the paradoxical nature of scientific communication during the Cold War,

FOREWORD

where astronomers aimed to communicate with extraterrestrials while struggling to bridge the communication gap between their own nations. In this light, we will find that CETI was not such a fringe science after all, but a phenomenon that shaped the ideas, technology, and culture of astronomy in the mid twentieth century. In 1977, engineer and CETI scientist Barney Oliver wrote: "It is easy to dismiss [CETI] as romantic, chauvinistic nonsense, but is it? We suggest that it is chauvinistic and romantic but that it may not be nonsense."[3] The contents of this book are also chauvinistic and romantic, but I hope to show that they are not nonsense.

Before beginning, I would like to take a moment to address the laundry list of incredible, brilliant, and kind people who have supported me and this project over the years. First, I thank my family, especially my handsome, lovely, sweet, and wonderful husband, Jim, whose support and encouragement were invaluable over the past few years. That being said... I'd have finished writing this book much more quickly if it weren't for you!

I owe my career to Ken Kellermann, emeritus scientist and historian at NRAO, and also to Ellen Bouton, archivist at NRAO. Their support, wisdom, and kindness changed the course of my life and influenced me deeply. I am also immensely grateful for the support of the National Radio Astronomy Observatory, an institution that has assisted my research for many years and in countless ways. I would especially like to thank Director Tony Beasley and Assistant Director Tony Remijan, but also Tristan Ashton (and Faye!), Erica Behrens, J. J. Burns, Barry Clark, Sophia Dagnello, Chris De Pree (and Sheryl!), Cosima Eibensteiner, Aaron Evans, Miller Goss, Nan Janney, Amanda Kepley, Erica Keller (and Allyson!), Amy Kimball, Anna Kapinska, Pearl Lara, Ryan Loomis (and Diana!), Andy Lipnicky (and Camille!), Dominic Ludovici, Brett McGuire, Anthony Pearson, Scott Ransom, Michael Sanchez, Samantha Scibelli, Haley Scolati, Jackie Villadsen, Cole Wampler (and Ashlyn!), Akeem Wells, Sarah Wood, and every other person who helped me during my time at NRAO. Go Janskys!

I also thank my colleagues from my time at the Harvard and Smithsonian Center for Astrophysics, especially Daina Bouquin, Katie Frey, Giancarlo Romeo, Emily Margolis, Charles Alcock, and Urmila Chadayammuri.

I thank the historians and other social scientists who trained, influenced, and supported me, especially my PhD supervisor, Richard Staley, and my PhD advisor, Simon Schaffer, as well as Robert Illife, Steven Dick, David DeVorkin, Kathryn Denning, Woody Sullivan, William Barry, Sloan Mahone, Erica Charters, and Stephen Garber.

I offer sincere thanks and warmth to the Drake family, especially Frank, Nadia, and Spencer (the future historian extraordinaire!). I also thank the rest of the SETI community, especially Jason Wright, Jill Tarter, the members of the St. Andrews SETI Post-Detection Hub, and the members of the Order of the Octopus.

Finally, I offer thanks and solidarity to my Ukrainian, Russian, and former Soviet colleagues, especially Sasha Plavin, Margarita Drochkova, Yuri Kovalev, and most especially Leonid Gurvits. Your bravery, wisdom, activism, and kindness in the face of treachery continue to inspire.

Introduction

How can we interpret signals from extraterrestrials when we struggle to interpret signals on Earth? We will begin not in space, but at the bottom of the ocean.

> October 27, 1962 – the Soviet Foxtrot-class submarine B-59 cuts quietly through the depths of the Caribbean Ocean, armed with a nuclear-tipped torpedo. Above the surface, the Cuban Missile Crisis is reaching its height.
>
> Suddenly, explosions. Left and right, the submarine is shaken by depth charges dropped by a US ship overhead. Concerned that they are under attack, and perhaps that nuclear war has broken out above the surface, the captain, Valentin Grigorievitch Savitsky, decides to fire the sub's nuclear torpedo. To do so, he must first have unanimous support from two other officers, the political officer and the deputy.
>
> President Kennedy had earlier declared a blockade of sea traffic between Cuba and the United States. Robert McNamara, the US secretary of defense, sent to Moscow and Soviet submarines radio messages headed "Submarine Surfacing and Identification Procedures," which stated that the US Navy would take action to "induce [submarines] to surface and identify themselves" if found violating the blockade. The US Navy had orders – not to attack Soviets, but to drop warning charges to prompt submarines to surface.[1]
>
> B-59 did not hear McNamara's message; it was too deep to receive radio communications.
>
> The political officer, Ivan Semonovich Maslennikov, gave his authorization to fire. But the final officer, Deputy Vasili Arkhipov, refused to authorize. According to the ship's communications intelligence officer, Arkhipov did not see the charges, which had been dropped only to the sides of the submarine, as a hostile act of war.
>
> This is not an attack, he argued. This is a signal.[2]

The history of the Cold War is in large part a history of signals: signals from the ocean, signals from the Earth, and signals from outer space. Unlike the two previous world wars – which were fought in trenches, boats, and airplanes, using toxic gases, guns, and bombs – many of the most significant features of the Cold War were battled through adherence to a set of signaling and listening practices and variations on them, including intelligence-gathering masked as diplomacy, satellites peeping from overhead, and scientific progress disguising threats of destruction. Fundamentally, signaling is a form of communication, and communication was the battlefield of the Cold War. The above episode also demonstrates that communication is rarely straightforward, especially between cultures that are foreign – or alien – to each other. Cold War communication often relied on signals and codes, and therefore was rife with the potential for miscommunication. After all, were it not for Deputy Arkhipov's[a] understanding that the US charges were attempting to communicate a desire that the Soviets come to the surface and not a hostile act of war, the Soviets might have somewhat reasonably decided to retaliate with their nuclear arsenal, potentially igniting a "cold" war into a hot conflict that could lead to global annihilation.

In addition to signals, the history of the Cold War also concerns aliens. Cold War-era science fiction, spurred by the Space Race and by fears of an attack from above that were in turn prompted by the launch of Sputnik (and the beep-beep of the signal it transmitted) and by the rise of atomic weaponry, foretold alien invasions and first contact scenarios with a combination of delight and terror. Like the entire Cold War, these science fiction stories of extraterrestrials fundamentally concerned communication with foreign cultures – the act of sending, listening to,

---

[a] Note on the transliteration of Russian names and words: I have chosen to transliterate Russian in a way that reflects the original Cyrillic spelling. For example, I would choose to transliterate the name Геннадий Шоломицкий as Gennadii Sholomitskii, representing ий with a double *i* as opposed to a single *y* and using a double *n* instead of a single *n*, even though this does not alter the pronunciation. I have chosen this transliteration style to aid future scholars in the field who may want to search Russian and Soviet sources: they should be able to reverse-engineer my English spellings into Russian easily. An exception is the name of Iosif Samuelovich Shklovsky, which under my style rules would be transliterated "Shklovskii" but whose bearer personally preferred a transliteration with *y*. Because this book owes a great deal to Shklovsky, I have decided to deviate from my style to honor his wishes.

and interpreting signals. Take for example Gene Roddenberry's television show *Star Trek*, which first aired in 1966. *Star Trek* is set in the twenty-third century on the starship *Enterprise*, a military–scientific vessel operated under the auspices of the United Federation of Planets and tasked with both "exploring strange new worlds" from a scientific–technical perspective and maintaining peace throughout the galaxy through military power. Take away the rubber costumes and planet-hopping and *Star Trek* is simply a show about Cold War international relations and the scientific–military–industrial complex.[3] As with the Cold War itself, much of the show concerned signals. *Enterprise* was constantly in communication with alien civilizations, picking up distress signals and being tasked with the difficult challenge of making first contact.

A spin-off from the original series, *Star Trek: The Next Generation* had an episode focused entirely on the challenges of communicating with the alien. The episode "Darmok" began with the captain of the *Enterprise*, Jean-Luc Picard, becoming marooned on an alien planet inhabited by a species whose members could communicate only in metaphors drawn from their own complex mythology.[4] The entire 40-minute episode is dedicated to Picard's frustrations as he tried to conduct meaningful communication with a people whose culture and way of signaling differed significantly from his own. The act of communication with the "other" is a major theme of Cold War and post-Cold War science fiction precisely because it parallels the attempts of communicating with the "alien" on our own planet – those nations and peoples whose languages, cultures, and ontologies differ drastically from the familiar ones.

There was also an underlying sense of anxiety during the Cold War, similarly evoked in alien science fiction. Film theorist Susan Sontag, in an essay on science fiction evocatively titled "The Imagination of Disaster," famously wrote:

> Here is a historically specifiable twist which intensifies the anxiety. I mean, the trauma suffered by everyone in the middle of the 20th century when it became clear that, from now on to the end of human history, every person would spend his individual life under the threat not only of individual death, which is certain, but of something almost insupportable psychologically – collective incineration and extinction which could come at any time, virtually without warning.[5]

But why was science fiction so prominent a genre in the United States and in the Soviet Union during the Cold War period? The rise of science fiction can likely be explained by the apotheosis of science in the postwar period. The Cold War's scientific–technical competition, arms race, and Space Race elevated the status of science and technology in both countries. Furthermore, one of the primary battlefields of the Cold War was outer space – not only in the Space Race, but in the less public race for gathering intelligence with the help of satellites and techniques of signals intelligence. As seen in *Star Trek*, aliens were a convenient stand-in for foreign civilizations with which we struggled to communicate and build some understanding. Considering this combination of science adulation, xenophobia, and newfound public awareness of outer space, it is no wonder that science fiction became a primary medium through which to express Cold War anxiety and aliens became its mode of expression. The subject of this book is not science fiction, yet I begin with this brief analysis to highlight the interconnected nature of science, warfare, anxiety, and aliens in the Cold War mindset. Recognizing these connections, let us turn our focus on the development of two sciences that arose in the early Cold War period: radio astronomy; and communication with (and search for) extraterrestrial intelligence.

A Cold War Science

Radio astronomy is a subdiscipline of astronomy that observes the universe in the radio part of the electromagnetic spectrum. Instead of the mirrors and lenses that optical telescopes use to observe the "visible" part of the spectrum (meaning those wavelengths of light that human eyes can perceive), radio telescopes use receivers, parabolic dishes, feed horns, and antennas to explore the "invisible" universe (meaning light in the cosmos that has a wavelength too long for the human eye to detect). Despite what you might have seen in Netflix's *Three Body Problem* and in other science fiction shows, radio astronomers do not use headphones to "listen" to radio signals; radio waves are not sound waves, they are light that our eyes can't detect. Radio astronomers use creative ways to visualize radio waves, often in the form of spectra and data-processing techniques that convert raw radio wave data into meaningful images and maps.

INTRODUCTION

The rise of radio astronomy as a scientific discipline was not simply a revolution in astronomy but also a significant diplomatic development in the Cold War – integral to the facilitation of international scientific collaboration and citizen diplomacy during an otherwise geopolitically contentious period. Not only was radio astronomy an intrinsic part of mid twentieth-century scientific research in both the United States and the Soviet Union, but it helped promote a philosophy of scientific internationalism and facilitated successful scientific exchanges between nations locked in conflict. The reasons for these successes are twofold.

First, Cold War-era radio astronomers developed scientific techniques and experiments that sometimes necessitated global cooperation, incentivizing scientists in the United States and Soviet Union to circumnavigate political barriers to achieve their science goals. Their "science first" approach resulted in many successful collaborative experiments in the 1960s and 1970s that led to the development of long-standing partnerships between research groups in the United States and the Soviet Union, then Russia, culminating in cooperation on contemporary projects and missions such as RadioAstron, a space-based radio interferometer. As part of my field work, I traveled to Moscow to conduct interviews with scientists in attendance at the RadioAstron International Science Council, held at Russia's Astro Space Center in October 2019. As an example of a more recent large collaborative project between US and formerly Soviet astronomers, I was curious to see whether the internationalist mentalities of 1960s radio astronomy were still alive in the present day. One astronomer who had worked with colleagues in the Soviet Union during the Cold War told me:

> This is one of the major strengths of an area like radio astronomy . . . because you don't just deal with your own country, you're literally dealing with the world . . . The Cold War was going on but the scientists were still collaborating . . . these things are separate from politics, and I think that's a very powerful phenomenon.[6]

Interestingly, that interview was conducted on October 4, 2019 – the 62nd anniversary of the launch of Sputnik, the Soviet satellite that became the first human-made object to orbit the planet and gave the Space Race an extraordinary public profile.

Sputnik and the subsequent Space Race are key examples of how science and technology can be politically charged and subject to interference determined by non-scientific motivations. Fervent internationalism and cooperation aside, radio astronomy was no exception. In other words, despite best efforts when "dealing with the world," it was impossible to avoid worldly challenges. Cold War-era radio astronomers collaborating with one another from either side of the Iron Curtain faced many problems that stemmed from geopolitical conflict, such as travel bans, mail interference, inconsistency in data sharing, and obtrusion from the intelligence community. Therefore the second reason for the unusual collaborative success of radio astronomy during the Cold War period was the development of a unifying philosophy that transcended political boundaries. Out of radio astronomy grew a subfield: communication with extraterrestrial intelligence, CETI, which fostered an unprecedented level of international collaboration focused on a common quest that was seen as universally significant, not nationalistic.

CETI has long been a part of the infrastructure and scholarship of radio astronomy. Some contemporary radio astronomers are resistant to this view, as CETI has been viewed by some as existing on the fringe of radio astronomy. At best, it was a waste of telescope time; at worst, a damaging pseudoscience. Scientists who became involved in CETI sometimes did so at the risk of their careers. But CETI was (and remains) a foundational subfield within radio astronomy, one that contributes to building and shaping networks of cooperation even during unusually hostile geopolitical times.

CETI's unique collaborative success resulted from the general philosophy of the community. In considering the potential cultural impact of discovering extraterrestrial intelligence (ETI), prominent CETI scientists such as Carl Sagan and Frank Drake argued that the discovery of life on other worlds might bring about global unity. They strove to cooperate with their global peers as "earthlings," not national citizens.[7] Because of this cosmopolitan perspective, CETI assisted in the formation of networks of contact and communication between Soviet and American[b] astrophysicists, which led to collaboration in other areas of

---

[b] The use of the term "American" to refer to citizens of the United States is sometimes contested, as some argue that it erases the identities of people who live in "the Americas," that is, the

radio astronomy, regardless of political challenges. Yet despite its success in promoting international collaboration and camaraderie, there was another side to Cold War radio astronomy and CETI – their roles in supporting, and sometimes appropriation by, militaries and governments.

During the Cold War, the United States considered scientific freedom an instrument of warfare.[8] This framing was due in part to the ideological campaigns waged by the United States and the Soviet Union in their fight for global dominance. In the Soviet Union, the demand that science serve the interests of the people (and state) involved censorship of scientific literature,[9] state-mandated theories,[10] adherence to Marxism,[11] and (in extreme cases)[12] the murder and imprisonment of scientists whose actions or beliefs did not align with the politics of the day. In the United States, the ideological campaign was far more subtle and arguably more insidious. To present itself as a foil to the Soviet Union, the United States concocted an ideology of "scientific freedom," which was in practice synonymous with its brand of free-market capitalism, democracy, and individual liberty.[13] American science was presented as apolitical; unlike their Soviet peers, American scientists did not have shadowy government censors like Glavlit reviewing their work, nor did they face travel bans for not joining a given political party. American scientists were free to disagree with each other's theories in independently published scientific journals, mostly without fear of imprisonment.[c] Yet this presentation of "scientific freedom" as a tool to promote democratic values to "unaligned" countries masked another side of American science – a side that was not apolitical but deeply entrenched in politics, the military, and imperialism. The notion that science promoted democracy by providing a rational framework for the pursuit of truth was undermined by the reality that science in the United States was co-opted by the military–industrial complex; there was essentially no field of science in the mid twentieth century that was untouched by the influence of the US military. Furthermore,

---

continents of North and South America, but not the United States. That said, given that the main bulk of this book focuses on the period between 1950 and 1980, when "American" referred primarily to citizens of the United States, to avoid confusion, I have chosen to use this adjective here just as the historical figures I write about did (and would have done in my place, too).

[c] Of course, there is a history of the United States impugning scientists for ideological reasons, as happened when the US government removed the security clearance of J. Robert Oppenheimer, known as "the father of the atomic bomb," when he was accused of being a communist.

US scientific freedom disguised US imperialism, which directly benefited the construction of scientific facilities and instruments on settled land, as part of its scientific–technical competition with the Soviet Union. With this understanding, science and its institutions seem to be central agents of democracy, and yet they look potentially exploitative and tyrannical.

Radio astronomy and CETI are revealing case studies within the larger history of Cold War science because of the manner in which these specific branches of science embodied the dual tensions of Cold War scientific institutions. They played pivotal roles in promoting internationalism and scientific freedom while simultaneously being implicated in two central ills of the Cold War: espionage and nuclear proliferation. As we shall see, by virtue of their scientific goals, CETI radio astronomers became adept at developing tools and techniques that aimed to target and identify intelligent extraterrestrial signals from space. This made the field ripe for exploitation by the intelligence community, which used the signal detection and analysis techniques developed by CETI to enhance its deep-space surveillance capabilities.[14] These instruments were also used for missile tracking and nuclear defense systems. Such confluences shaped the ideas of CETI; as CETI scientists speculated on the evolution of technological civilizations in the galaxy, they wrestled with the future of their own technological civilization, as it faced the new threat of self-annihilation. Clearly, historical investigations into the political and social hurdles faced by radio astronomers and their international and regional communities during the Cold War undermine the belief that science was separate from politics and demonstrate that internationally cooperative sciences bear worldly challenges in addition to scientific ones.

The Cosmic Mirror

This is a history about a science, about a war, and about a question; but fundamentally this is, at its very core, a history of communication. The present book rests on a small irony: although they were focused on communicating with extraterrestrials, astronomers in the Soviet Union and the United States were also quite alien to one another, and communication across the Iron Curtain was in some sense just as much of a challenge as communication across the depths of space. This dynamic should not surprise us, as the United States and the Soviet Union have long alienated each other.

Consider the meeting of US and Soviet troops at the end of World War II. On April 25, 1945, the Soviet Red Army and the US Infantry successfully cut the German army in two at the Elbe River, southwest of Berlin. The day is still informally celebrated, and sometimes called a "first contact" between the United States and Soviet forces. The term "first contact" is an appropriate one, for the soldiers on either side of the river appeared to expect aliens rather than fellow human beings. Luibov Kozinchenko, a Soviet soldier from the Red Army's 58th Guards Division, later recalled the day: as the Americans crossed the river, "[w]e could see their faces. They looked like ordinary people. We had imagined something different."[15] On the US side, Al Aronson, an American soldier from the US 69th Infantry Division, claimed: "I guess we didn't know what to expect from the Russians. But when you looked at them and examined them, well, you could put an American uniform on them and they could have been American!"[16] Both sides appeared surprised at the humanness of these people, who had until then seemed quite alien.

This book imagines four potential audiences – historians, astronomers, SETI researchers, and a curious general public – and envisions they will each take away a unique perspective from their reading. Historians will gain C/SETI as a novel entry point into our understanding of the complex cultural and political dimensions of Cold War science and technology. Astronomers, I hope, will learn to see C/SETI not as a fringe pseudoscience, but as a field that shaped the structural and cultural foundations of their discipline. SETI researchers will learn new perspectives on the history of their field, especially the oft neglected Soviet side, and will also gain an appreciation for the fact that including interdisciplinary perspectives in the search for intelligence in the universe only stands to improve the search. Ultimately I wish for all my readers to appreciate the creativity, tenacity, and optimism of Cold War-era CETI scientists, especially those from the former Soviet Union (such as the Ukrainian Soviet astronomer I. S. Shklovsky, whose scientific accomplishments and feats of bravery have not yet fully reached the Western world).

This book pretends to be about aliens, but it fundamentally concerns human beings and two of their primary emotions: hope and fear. As one SETI scientist once put it to me, "science is not only hard because science is hard; it is hard because we live in the world."[17] This is a complex subject. CETI both undermined the geopolitical goals of nation states through international collaboration and actively supported militaries through assistance in intelligence gathering. CETI scientists alienated fellow earthlings at the same time as they strove to depict our world equitably in messages to extraterrestrials. CETI inspired antinuclear activism as well as aided in developing tools and techniques that made nuclear strikes more accurate. It simply isn't possible to conclude that radio astronomy and CETI were definitively forces for either world peace or global oppression. Instead, we find ourselves sat within the uncomfortable tensions of this history, and we can use CETI as a vessel through which to analyze the contradictions and complications of Cold War science and philosophy.

The Cold War-era aliens of *Star Trek* were human beings dressed up in costumes with the addition of pointy ears, green skin paint, or cybernetic implants; nevertheless, they were highly anthropomorphic. It is now a common trope and a refrain in SETI that we must avoid anthropocentrism, the tendency to interpret or project homocentric values and characteristics onto our expectations of extraterrestrial life. Some have argued that anthropocentrism is unavoidable, while others have argued that being cognizant of our tendencies toward anthropocentrism can foster creative, out-of-the-box thinking. It is not in any sense a new idea to point out that human culture and bias infiltrates our imaginings of extraterrestrial intelligence.

In my view, however, the thesis "C/SETI is anthropocentric" reflects an incomplete thought. Certainly, you will find many instances of anthropocentric (and even Eurocentric or male-centric) thinking in the pages of this book. But anthropomorphism here is an asset, not a limitation. In this book, our task is to understand not the nature of the alien, but human beings and their creations, including science and war. Our biases, chauvinism, and motivations shape our world, and in studying them we gain a novel perspective on, and way into, our own history.

Jill Tarter, a SETI scientist and prominent public advocate for the search, defined a theory she has called "the cosmic mirror"[18] (Figure i.1). Tarter has argued that the pursuit of C/SETI is valuable even if no evidence of extraterrestrial intelligence is ever detected, because the

INTRODUCTION

Figure i.1 The Cosmic Mirror – In art history, the motif of the mirror has long served as a device to chastise the viewer's (often a woman) sense of vanity. The original painting by Caravaggio shows the sisters Martha and Mary from the New Testament. Martha is trying to convert Mary from her hedonistic life to a more virtuous life following Jesus. In the original painting, which shows a convex mirror, uses the mirror to symbolize the life of self-absorption she is about to give up. In this edited image, however, the Cosmic Mirror replaces a conventional mirror, reminding the viewer of our small place in a vast cosmos and inviting contemplation of humanity's interconnectedness. This reinterpretation shifts the focus from individual self-preoccupation to a broader reflection on our place within the cosmic order, inviting viewers to ponder deeper existential questions. Instead of being an object of vanity to be cast aside, the mirror transforms into an object that encourages us to seek deeper meaning.

introspection and unity it prompts are valuable in and of themselves. She describes the cosmic mirror as

> the mirror in which all humans can see themselves as the same, when compared to the extraterrestrial 'other'. It's the mirror that allows us to alter our daily perspectives and see ourselves in a more cosmic setting. It is the mirror that reminds us of our common origins in stardust.[19]

With the cosmic mirror, Tarter argues that C/SETI reflects a universalizing image of humanity, juxtaposed against the truly alien. I wish to take her definition a short step further – into what I will later label and define as the cosmic prism – and argue that CETI reveals not a universal portrait, but the many-shaded schisms, conflicts, and contradictions of our world. It unveils the locality of our instruments of both science and warfare, our cultures and ideologies, and our Earth-bound and orbital systems of communication. Perhaps no other science shows us as much about ourselves, and prompts us to consider: do we like what we see?

1

What Is CETI?
Reframing the History of Communication with Extraterrestrial Intelligence

I want to begin by asking "just what *is* CETI?" and, perhaps more critically, "what is it *not*?" Simple enough questions, but ones with exceedingly complex and contradictory answers. Most people, when they consider the scientific efforts to discover the existence of extraterrestrial intelligence, will use the term "SETI," which is an acronym for "search for extraterrestrial intelligence." Many would immediately think of the SETI Institute, a scientific organization founded in 1984 that funds research and education on the search for intelligent life in the universe. They may think of prominent SETI scientists, such as Seth Shostak or Jill Tarter. The exchanges between Ellie Arroway and Palmer Joss from the hit science fiction film *Contact* (1997) may come to mind. But SETI is a relatively new term; it only entered use in the scientific community in the early 1970s and did not gain wide traction before the 1980s. In the 1960s, when astronomers started to use radio astronomy equipment to conduct searches for evidence of extraterrestrial signals, they tellingly preferred the term "CETI" – *communication* with extraterrestrial intelligence.

It was not until 1971, at a joint US–USSR conference on communication with ETI, that the shift from CETI to SETI slowly began, marking a shift in how the field was defined.[a] The 1971 conference was enthusiastically

---

[a] While the conversations at the 1971 conference marked the beginning of the end of CETI, in a documentary sense, the shift from CETI to SETI was decisively made during 1975–1976 Morrison workshops, whose proceedings explicitly state that Americans adopted the nomenclature "SETI" at that workshop to distinguish it from CETI: "The Soviets have named their program CETI or Communication with Extraterrestrial Intelligence. The acronym SETI (Search for Extraterrestrial Intelligence) was adopted by the Workshop and by the [NASA] Ames Research Center to differentiate our own efforts from those of the Soviet Union and to emphasize the search aspects of the proposed program" (Morrison, P., Billingham, J., and Wolfe, J., eds., *The Search for Extraterrestrial Intelligence: SETI*. National Aeronautics and Space Administration, Scientific and Technical Information Office, 1977, 213). This landmark report

centered on CETI with a "C." This term suited the conversations held at the conference; one attendee, Marvin Minsky – a computing pioneer in the field of artificial intelligence (AI) – presented a paper on how to use AI to communicate with extraterrestrial intelligence without the need for long response periods: simply launch and place intelligent computers around planets that might harbor intelligent life and let them do the talking for us.[1] Carl Sagan opened his conference remarks by noting how fitting the term CETI was for the field, because in Latin *ceti* is the genitive of the noun *cetus*, "whale," which he notes is "undoubtably another intelligent species inhabiting our planet," and also because the first CETI project, Project Ozma, observed a star system named Tau Ceti.

But it is generally recognized that a conversation took place at the conference that argued for transitioning CETI to SETI.[2] The conversation allegedly took place between Carl Sagan, American radio astronomer Frank Drake, and Soviet radio astronomer I. S. Shklovsky. According to oral history accounts, the group realized that scientists must first *find* extraterrestrial intelligence before attempting communication, and therefore a focus on "search" rather than "communication" would be more representative of the field's goals.[3] But there is another reason why the shift might have occurred. During the Cold War period, CETI scientists from the United States and the Soviet Union experienced immense difficulty in communicating with each other, because of cultural, political, and disciplinary differences. These difficulties led them to the realization that establishing effective communication with extraterrestrial intelligence could be an overwhelmingly challenging task. Indeed, upon leaving the Soviet–American CETI conference, Frank Drake remarked that he and his American colleagues returned home "having gotten a hint of what life on another world was like."[4] Perhaps in recognizing the various challenges of communication, both terrestrial and extraterrestrial, the Soviet–American cohort decided that the first problem to solve would simply be detecting evidence of extraterrestrial intelligence.

---

was therefore titled "The Search for Extraterrestrial Intelligence." Thus the Americans were trying to distinguish themselves from the Soviets, but unfortunately this has tainted our historical memory. Today many SETI scientists believe that "CETI" refers exclusively to the Soviet effort, even though both American and Soviet astronomers used "CETI" to describe their efforts before 1975 and the term "SETI" eventually took off in the Soviet Union, just as it did in the United States.

WHAT IS CETI?

This book will cover themes and histories that can sometimes be applied to SETI, but our focus will remain on the period between the 1950s and 1980s. CETI with a C was a distinctive phenomenon in the larger history of the extraterrestrial life debate and of the contemporary search for extraterrestrial intelligence, though it was connected to it. I choose to put parameters on CETI with a C because CETI was the term in use at the time when the field developed, and the choice to emphasize communication (as opposed to search) reveals important characteristics of the young science that tie it closely to the Cold War environment from which it emerged.

It is also important to define disciplinary boundaries. This task presents a rather difficult challenge, given the disciplinary shifts in the field over the past half century. When scientific CETI developed in the late 1950s, it was initially a subfield of astronomy with strong ties to radio astronomy. Scientists involved in CETI projects came from a variety of disciplinary backgrounds, but much of the early work in the field came from radio astronomers and their observatories and was published in traditional astronomical journals. Today the science involved in seeking out intelligent extraterrestrial signals has undergone a shift in its disciplinary grounding; for example, the Ad Hoc Committee on SETI Nomenclature, which convened in March 2018, defined SETI thus:

> n. A subfield of astrobiology focused on searching for signs of non-human technology or technological life beyond Earth. The theory and practice of searching for extraterrestrial technology or technosignatures.[5]

In the mid twentieth century, CETI certainly had some relationship with the newly developing discipline of astrobiology but remained largely rooted in the radio astronomy community.

* * *

Every chapter of this book is an attempt to persuade you, the reader, that CETI with a C was unique to the Cold War. Was it connected to the larger history of the extraterrestrial life debate? Certainly. Did many of the historical actors in the field later go on to identify as SETI scientists? Absolutely. But when CETI developed as a field, between 1959 and the early 1970s, it was so entangled in the Cold War that to attempt to understand it without giving considerable attention to the Cold War would be to completely misunderstand the young field. In this book I will use the term SETI when I refer to contemporary research on extraterrestrial intelligence and the term CETI when I refer to pre-1980 research (e.g. "I presented my research on Cold War CETI at a SETI conference in 2022"). When referring to subjects that apply to both CETI and SETI, I will use the term C/SETI.

My emphasis on CETI also reflects the theme of communication with the alien – in space and on Earth. I will use the term "alien" as a rhetorical tool, to highlight how US astronomers' attempts to communicate with their Soviet peers paralleled their attempts to communicate with extraterrestrial civilizations: in a sense, both were attempts at communicating with an unknown "other," be it extra- or inter-terrestrially. The goal is not to "alienate" people and cultures, but rather to highlight the distinctly human nature of the attempt to communicate with aliens. After all, for a book that is fundamentally concerned with the search for and the communication with extraterrestrial civilizations, I will devote considerable attention to civilization on Earth – analyzing our militaries, our scientific infrastructures, and the frameworks of our international relations. In fact, for much of Chapter 2, CETI will hardly be discussed at all. This is because I believe that the history of CETI can be best understood as a history of "humanness" – in other words as revealing splits and bifurcations between the better and the worse features of humanity.

In this chapter I will give a brief history of history; historians call it a "historiography." Historiographies can sometimes be dull accounts of a field and are often left out of history books altogether. It is important for history students to address historiography in their essays and dissertations; it demonstrates the work they have put into understanding the background of their field, and it shows that they appreciate the context of their research. But it is assumed that the author already understands the historiography of the field on which they are writing. Therefore, when writing for a general or interdisciplinary audience, historians are generally advised to relegate historiography to endnotes in their books and avoid boring their readership or coming across as pedantic. The Organization of American Historians warns us: "General readers prefer narrative ... The goal is ... not to demonstrate your knowledge of historiography or historical debates."[6] This book will choose to reject this sound advice entirely – but not without good reason. The reason why I chose to include this chapter is that I believe that CETI's historiography is particularly significant for our understanding of the science. CETI is an unusual field; it is inherently interdisciplinary, it is difficult to put parameters on, and it deals with incredibly abstract questions. Because of this, historians as well as physical scientists have disagreed on how to contextualize the science, or even on deciding whether it is indeed a science at all.

New Technologies and the Possibility of Strange and Exotic Beings

Sometimes in science the ideas precede the technology. Physicists, for example, will often build specific instruments to test their theories, as Earnest O. Lawrence did when he invented the cyclotron to confirm his idea that he could use magnetic fields to accelerate charged particles. The Large Hadron Collider was specifically built to address several fundamental questions in particle physics, such as the search for the Higgs boson particle (and in 2012 it did discover the Higgs).

In the history of radio astronomy, the opposite has often been the case: new developments in technology have driven the scientific questions and discoveries. Many of the discoveries in radio astronomy have been serendipitous, arrived at by accident, when instruments were used for other purposes.[7] During World War II radar operators in Britain made surprising discoveries; the detection of radio waves emanating from the Sun is one example. The discovery of pulsars (highly magnetized, rotating neutron stars) was made when then graduate student Jocelyn Bell noticed a strange series of regular pulses in her observations of quasars (she and her advisor initially called these mysterious pulses "LGMs" – little green men). Even Karl Jansky, who is generally considered the first radio astronomer, made his discovery in a largely accidental manner. In 1933 he was working for Bell Labs, trying to determine the cause of the background static that plagued telephone communications.[b] In attempting to find the source of the static, which initially he believed might be caused by lightning, he ultimately discovered that the strange "hiss" was coming from no other place than the center of our Milky Way galaxy.

---

[b] You see, radio astronomy has been connected to communication from the very start!

All these discoveries were possible because the development of new tools and technologies opened new windows into the universe. Like with radio astronomy, the development of CETI arose from new technologies – and it is all thanks to none other than the radio.

It is no coincidence that the dawn of radio technology led to questions about communication with extraterrestrial life. The newfound ability to communicate with foreigners across the expanse of the sea prompted speculation that the same tools could be used to communicate across the expanse of space. In 1896, Serbian American electrical engineer and physicist Nikola Tesla asserted that his new electrical transmission system could be used to communicate with Mars, the Red Planet; and, soon after, Italian radio engineer Guglielmo Marconi claimed to have received radio signals from it. Although Tesla alleged to have used his radio experiments to become "the first to hear the greeting of one planet to another" in 1901, it was not until about two decades later that using radio to communicate with other planets became widely discussed in scientific circles.[8]

In January 1919, the *New York Times* published an article titled "Radio to Stars, Marconi's Hope."[9] It contained a summation of an interview the English journalist Harold Begbie had conducted with Marconi during which the latter discussed "the possibility of communicating by wireless with the stars."[10] In the interview, Marconi speculated on the potential use of radio technology for interstellar communication. He said:

> Messages that I sent off ten years ago have not yet reached the nearest stars. When they arrive there why should they stop? . . . That is what makes me hope for a very big thing in the future . . . communication with intelligences on other stars.[11]

The very next year, Marconi announced that he was investigating signals he speculated might have come from Mars. The *New York Times* once again published a lengthy article on Marconi's so-called Mars signals.[12] In Europe, Marconi's claims faced mostly ridicule; one French newspaper publicized his discovery under the headline "Hello, Central, Give Me the Moon," referring sardonically to Marconi's wireless telegraphy system.[13] Yet the signals Marconi received, which he described as "distinct but unintelligible," generated much excitement in the United

WHAT IS CETI?

Figure 1.1 *The Tomahawk*, a Minnesota newspaper, published an article expounding on the various ideas and claims concerning extraterrestrial communication made by Edwardian engineers such as Nikola Tesla, Guigelmo Marconi, and Thomas Edison. Credit: *The Tomahawk*. White Earth, Becker County, MN, March 18, 1920. Chronicling America: Historic American Newspapers. Library of Congress.

States. A newspaper in Minnesota, *The Tomahawk*, published an article titled "Hello Earth! Hello!" that jumped at the possibility of communication with extraterrestrial intelligence (Figure 1.1). Marconi is quoted as having said:

> If there are any human beings on Mars I would not be surprised if they should find a means of communication with this planet. Linking of the science of astronomy with that of electricity may bring about almost anything.[14]

Still, despite interest and international conversation, these early investigations into the use of radio technology to communicate with extraterrestrials were rooted in conjecture and imagination, not in rigorous scientific investigation. For example, according to *The Tomahawk*, Marconi proposed that Martians may use Morse code to communicate with Earth; and he described the possible occupants of the planet as "human beings" rather than "aliens," which demonstrates that, although ideas for interplanetary radio contact were forming, there was no serious scientific investigation into the problem of communication with extraterrestrial intelligence. If anything, there was a Christian undertone to many of these discussions; if humans were intelligent beings made in God's image, then we would of course expect to find the galaxy populated with "people," not with truly alien creatures.

Nevertheless, the radio connection is important. One might wonder why CETI emerged out of the development of radio astronomy rather than optical astronomy. After all, both sciences do essentially the same thing: they observe light in the universe. Yet scientific CETI did not arise until the late 1950s and early 1960s, at the onset of radio astronomy's transformation into a scientific discipline. CETI's development out of radio astronomy, as opposed to optical astronomy, was due to the close association of radio technology with communication, intelligence gathering, and warfare. In other words, it was the detection of unknown signals in space – initially unexpected, because they originated from the military – that led scientists to speculate about the possibility that truly alien signals might also exist.

Many early radio astronomers began their careers as radio technicians who served in the military. This was true of the aforementioned Frank Drake, a radio engineer for the navy who pursued a doctoral degree in astronomy at Harvard University in 1955, after having completed his service.[15] Given his background, Drake became one of the first graduate students to complete the new program in radio astronomy at the university. On one night during his studies, Drake was observing the Pleiades star cluster using the university's 60-foot radio telescope. During this observation, Drake detected "what appeared [to him] to be an intelligent signal from an extraterrestrial civilization" coming from the cluster. He assumed its artificiality on account of its striking regularity (something that, at that time, had not been observed in the natural radio).[c] In his autobiography, Drake recalled the event as being so shocking to him that it very probably explained why his hair turned prematurely white

---

[c] Interestingly, about a decade later, in 1967, University of Cambridge graduate student Jocelyn Bell Burnell would discover what she called "LGMs" – little green men – later labeled "Pulsars." Pulsars are highly magnetized neutron stars that discharge beams of radio emission while spinning on their axis, which results in a strikingly regular "pulse" signal if observed from Earth. Mistaking new, previously unobserved phenomena for evidence of extraterrestrial intelligence is a long-standing pattern in the history of astronomy.

Figure 1.2 Frank Drake, pictured here at the age of 32, shortly after completing Project Ozma at the National Radio Astronomy Observatory. Drake jokingly credited the appearance of his prematurely white hair to the shock of believing that he had found evidence of extraterrestrial intelligence.
Credit: NRAO/AUI/NSF.

shortly after, although he was only in his twenties (Figure 1.2). He tested the signal's location by moving the telescope off the cluster; if the signal disappeared and then reappeared when he returned the telescope to the cluster, it would verify that the signal was indeed coming from the Pleiades. Unfortunately, Drake recalled that, to his "great disappointment," the steady, artificial signal continued to broadcast even when he moved the telescope off the cluster, which Drake assumed meant that "it had to be some form of terrestrial interference, probably military."[16]

Figure 1.3 This grainy image shows the inside of the control room of the 40-foot telescope in Green Bank, West Virginia. I took this picture while on an observing run at the telescope, using a disposable film camera (electronics that give off radio waves, such as digital cameras, are not permitted around the instruments; see Chapter 2 for more on this). Although Drake used the 85-foot telescope for Project Ozma, the project's radiometer, pictured here, has now been repurposed for the 40-foot telescope. On the chalk board, "1420.41" indicates that observations are looking at the hydrogen line, just as Frank Drake used to do for Project Ozma (see Chapter 4 for more on the significance of the hydrogen line). Credit: Rebecca Charbonneau.

The event lit a fire under Drake, and he became obsessed with the idea of using radio telescopes to seek out artificial signals from extraterrestrial intelligence. Immediately after completing his PhD, Drake was hired at the newly established National Radio Astronomy Observatory (NRAO). In April 1960, shortly after taking up this position, Drake designed a receiver to fit the Observatory's 85-1 telescope and conducted what is generally considered the first scientific radio search for extraterrestrial intelligence (see also Figure 1.3). His search targeted two nearby star systems, Tau Ceti and Epsilon Eridani, in the hope of detecting another artificial signal – but one of extraterrestrial origin this time. He named his search "Project Ozma," after Princess Ozma from L. Frank Baum's Oz novels. His justification for the name was that Oz was "a land far away, difficult to reach, and populated by strange and exotic beings," perhaps not terribly different from the worlds he was trying to communicate with.[17] In the Oz novels, the narrator uses wireless radio technology to communicate with the distant world of Oz: "As you know, I am obliged to talk these matters over with Dorothy by means of the 'wireless,' for that is the only way I can communicate with the Land of Oz."[18] Like the narrator, Drake wished to use radio to speak with exotic worlds somewhere over the rainbow.[19]

Interestingly, the same series of events that happened to him at Harvard took place again. After observing Tau Ceti inconclusively, Drake and his two student assistants, Margaret Hurley and Ellen Gunderman, moved the beam of the 85-1 telescope towards the direction of Epsilon Eridani, where they immediately detected an artificial signal.[20] After a few minutes of excitement, Drake once again realized the signal was of Earth origin, which he speculated was perhaps also military in nature.[21]

When historians discuss Project Ozma, it is usually in the context of the shift in the status of the extraterrestrial life debate – the long history of human speculation on the question of whether life exists beyond Earth. The extraterrestrial life debate preceded CETI by thousands of years. In ancient Greece, atomists in particular theorized about the plurality of *kosmoi* and pondered on the philosophical consequences of a universe of many worlds.[22] In early modern Europe, the Dutch mathematician Christiaan Huygens famously published in 1698 a treatise titled *Cosmotheoros: or, Conjectures Concerning the Inhabitants of the Planets*, in which he speculated on the existence of other worlds inhabited by people, just like Earth was.[23] There were also early proposals for messaging extraterrestrials; the first known scientific proposition to establish communication with beings from other planets is commonly credited to Carl Friedrich Gauss, a German mathematician. In the early nineteenth century, Gauss proposed the construction of a colossal visual demonstration of the Pythagorean theorem, written on the Earth's surface and visible to beings who he believed might occupy the Moon.[24] Up until the mid twentieth century, however, there were few attempts at actually seeking out and communicating with these other beings, and none that were systematically scientific in nature or conducted by a professional scientist at a scientific institution. In consequence, much attention has been given to Drake's Ozma on account of its novelty, but also because it set off a buzz of interest within the scientific community, inspiring many subsequent searches.[25] It marked the start of C/SETI as we now recognize it, our attempt to seek out and communicate with intelligent life in the universe by using (primarily) radio telescopes.

Project Ozma and the start of radio CETI were clearly also the result of technology driving innovation, because the idea of using radio telescopes to search for extraterrestrial intelligence was a case of simultaneous invention. Around the time Drake was planning Ozma in 1959, two physicists at Cornell University, Giuseppe Cocconi and Philip Morrison, published a paper titled "Searching for Interstellar Communications" in the scientific journal *Nature*.[26] Like Drake, they too had realized that radio telescopes could be used for interstellar communication. Drake, Morrison, and Cocconi all came independently to the conclusion that searching the radio spectrum at the frequency of 1420 MHz (a transition line of neutral hydrogen)[d] would be the best choice for making intelligent contact.[27]

Like Drake, Morrison and Cocconi planned to conduct a search at that frequency and reached out to Sir Bernard Lovell, a British radio astronomer and the director of the Jodrell Bank Observatory in England, hoping to use the observatory's large Mark I telescope (now known as the Lovell telescope). Cocconi sent Lovell an itemized list of arguments in favor of CETI research that ended on a humble note, stating: "As I said before, all this is most probably fiction, but it would be most interesting if it were not."[28] Cocconi's choice to tie CETI to fiction was likely a poor one – while of course there were significant links between CETI and science fiction, this relationship often came to the detriment of CETI's legitimacy as a scientific pursuit. Unfortunately for Cocconi, Lovell was not an advocate for CETI and rejected his proposal, with the result that Drake, not Morrison and Cocconi, conducted the first scientific CETI project.

Nevertheless, Drake, Morrison, and Cocconi had developed the techniques and theories that were to drive the search for the next half century.

---

[d] The hydrogen line, sometimes called the 21-centimeter line, occurs when the spins of the proton and electron in a neutral hydrogen atom flip from being aligned to being opposite, releasing energy as a radio wave at 1420.4 MHz. This signal helps astronomers detect and map hydrogen gas in space. Drake, Morrison, and Cocconi all believed that, since the hydrogen line was important to radio astronomers, alien radio astronomers would naturally assume that Earth astronomers are observing at that frequency, and therefore wouldn't miss their message.

In fact, even today, contemporary SETI projects (e.g. Breakthrough Listen, launched in 2016) follow essentially the same methods as Project Ozma, although with far more sensitive and powerful instruments. Historians of the extraterrestrial life debate such as Steven J. Dick have argued for the significance of Project Ozma as launching a transformation in the way humans have approached extraterrestrial communication, one that has lasted up to the present day.[29] That being said, only a few historians have taken note of the specific connection between CETI and the Cold War. Nor has there been a comprehensive historical study of CETI, and especially not one that carefully examines CETI's development in the Soviet Union.

Instead, the existing literature on the history of CETI focuses largely on how CETI shifted the nature of the extraterrestrial life debate. In his 1999 book *The Biological Universe*, Dick (1999) briefly but correctly notes that the rise of CETI was the result of the development of radio astronomy, part of what he calls the "new astronomy," which recognized that observing the universe at wavelengths other than the optical was important for a holistic understanding of the cosmos.[30] Yet Dick did not examine the role of the Cold War in facilitating this development, choosing instead to describe Morrison and Cocconi as having "stumbled" into CETI "almost as an aside to their primary research."[31] Historians of radio astronomy, on the other hand, have certainly made the connection between the development of the science and the Cold War, especially Jon Agar, Woodruff Sullivan, and Kenneth Kellermann.[32] But, as Dick has pointed out, there is the "metahistoric issue of the cognitive status" of the extraterrestrial life debate and C/SETI; in layman's terms, historians struggle to agree on how to interpret and contextualize the long tradition of thinking about and seeking evidence of extraterrestrial life.[33] For example, historian Michael Crowe believes that philosophy was the main driver of the extraterrestrial life debate and argues that historical studies of the human preoccupation with extraterrestrial life should largely concern the human mind. He asserts:

> Although studying the history of ideas of extraterrestrial life may not shed light on such beings, it gives promise of telling humanity about itself. Just as inkblot tests are not about inkblots but rather tell us about their interpreters, just as a study of the paintings of saints may tell us little about those saints

but much about the artists and the era in which they painted, so also learning about how humans have thought of extraterrestrials can be deeply revealing of the fears and hopes of persons from the past as well as the images they have of the universe.[34]

Dick's study, on the other hand, shows how the extraterrestrial life debate was more than simply philosophy but was intimately connected to a wide span of scientific traditions, which he defines as an enterprise "composed of both philosophical and empirical elements" and encompassing atomist, Aristotelian, Copernican, and Newtonian worldviews.[35] Dick believes that, just as "the whole thrust of physical science since the seventeenth century scientific revolution has been to demonstrate the role of physical law in the universe," the extraterrestrial life debate within the scientific community has been driven by the desire to see if "an analogous biological law" exists in the universe.[36] Dick therefore situates the extraterrestrial life debate within the scientific community and sees CETI as stemming from twentieth-century scientific ideas and technologies; on the other hand, Crowe situates it within the history of ideas, not neglecting the role of science but principally believing that the pursuit was driven by philosophy.

Science or Pseudoscience?

I have explained how some historians have approached the history of CETI: most view it as a culmination of the extraterrestrial life debate, as having transitioned from a state of speculation into being a scientific field within the discipline of radio astronomy, and later astrobiology. Others, however, view CETI not as a science, but as a religious enterprise or as a pseudoscience, one that is perhaps not distinct from ufology. This idea was pursued by historian of technology George Basalla in 2006. Basalla, one of the few professional historians of science who have specifically addressed CETI in their research, makes the argument that CETI should not be considered primarily a scientific pursuit. In his 2006 book he argues:

> The idea of advanced extraterrestrial life emerged from a background of philosophical, religious, and scientific thought, with science the last ingredient

added to the mix ... Many modern scientists are not aware of the long and complex history, and the deep religious and emotional significance, of the idea of intelligent aliens. They are not dealing with scientific perceptions alone, but with old religious beliefs and philosophical concepts that underlie current scientific thinking.[37]

Basalla is a respected historian of technology, best known for *The Evolution of Technology* (1989), in which he posited a theory of technological evolution that emphasized the social, economic, and political contexts that affected that evolution. In *Civilized Life*, like Dick and Crowe, Basalla sees CETI as having grown out of the extraterrestrial life debate, and especially thanks to late Renaissance and Enlightenment thinkers, who needed to replace the empty void of space with supernatural beings. However, Basalla seems to view the religious themes in CETI thinking as incompatible with, or antithetical to, science: he frames CETI as a pseudoscience, and perhaps even as a religion in and of itself.

Basalla cites Drake's Christian fundamentalist upbringing as the source that "awakened his interest in extraterrestrial life," quoting Drake as stating: "A strong influence on me, and I think on a lot of SETI people, was the extensive exposure to fundamentalist religion."[38] But Basalla is uncharitable in his interpretation of statements from CETI scientists. If you examine the original quotation from Drake, which comes from a collection of oral history interviews taken by David Swift at a 1981 SETI conference in the Soviet Union, Drake continues:

> I learned enough of science and astronomy to recognize there was a conflict. Something was not jibing. Sunday school just did not fit in with what the universe seems to be ... there was a scientific basis for other alternatives. It wasn't just stuff written on old scrolls and things, which I was always suspicious of anyway. So to some extent it is a reaction to a firm religious upbringing.[39]

With this greater context, we can see that Drake's draw to CETI and astronomy was a reaction against religion. But Basalla believed that CETI scientists were entirely replacing religion with CETI, which he found just as religious in character.

Many contemporary SETI scientists take issue with this comparison with religion. When Basalla's book was first published in 2006, SETI scientists Jill Tarter and David Darling expressed their dissent on a SETI Institute radio program. Tarter explained: "SETI is an exploratory science, and we're simply trying to answer a question. If we knew what the answer was, and we're telling you what it was, then it would be religion."[40] Two of Basalla's major critiques of CETI were that, first, its proponents held a strong faith-like belief in intelligent life in the universe and, second, their attempts at contacting this life were rooted in anthropocentric assumptions about alien civilizations. To these points, Tarter retorted that belief in the likelihood of intelligent life in the universe was not inherently a religious idea. She claimed:

> I wouldn't work on a problem that I didn't think had a potential for a solution. So that I, I do in fact, think that what we know a little bit about[,] the environments in which life can exist on this planet, something[,] a very little[,] about how it originated, makes it plausible, that the same processes could have operated elsewhere. But I certainly wouldn't say I know anything about it, and what I believe or don't believe, doesn't matter.

In other words, it is a hypothesis with at least some empirical basis, not a belief rooted in religious faith. Regarding anthropomorphism, Tarter admits that anthropocentrism presents limitations in C/SETI, but not so many that it makes the science impossible. She argued: "we are stuck in our human skins. [We] can't get out of that. But rather than saying they're just like us, we are, in fact, restricted to saying that the technology that we're looking for, is that which we already understand."[41]

Of course, this is not to suggest there is no element of religion or spirituality in C/SETI. Religion has long been a part of the human experience and there are many scientists, C/SETI scientists and radio astronomers included, who hold deeply spiritual or religious beliefs. Furthermore, it is impossible to read works by C/SETI scientists without noting a religious tone that often underlies the scientific exposition. Basalla argues that Sagan was not aware of the "religious impulse that inspired [his] study of alien life." But this is clearly untrue: Sagan was perhaps especially interested in the interplay between religion and science. One particularly poignant example of this interest and awareness is Sagan's 2006 *The Varieties of Scientific Experience*. The title is a direct reference to William James' 1902 *The Varieties of Religious Experience*, which consists of lectures on the psychology of religion. The book is a collection of lectures given by Sagan that gave him, in the words of his wife Ann Druyan, a "chance to set down in detail his understanding between religion and science and something of his own search to understand the nature of the sacred."[42] Sagan also authored a science fiction book, *Contact* (1985), which later became a film and was centered on an imagined post-detection scenario. The tension between religion and science is inarguably the main theme in *Contact*. The lead characters are Dr. Ellie Arroway, an astrophysicist, and Palmer Joss, an evangelical preacher, and they have heated debates on the relationship between science and faith throughout the novel. *Contact*'s aliens often appear more like angels than extraterrestrials, coming to Dr. Arroway in the form of her beloved deceased father and preaching of universal love and community. The ending of the book hints at a larger, God-like construction of the universe through a hidden message in the number pi.

There is certainly a religious influence on C/SETI, as there is on many parts of science. Nearly every scientific enterprise has a connection to religion and religious institutions, and this is true throughout history. Indeed, many renowned scientists held religious beliefs that affected their scientific practice. The natural philosopher Isaac Newton, for example, was deeply religious. He believed his practice of studying the physical properties of the universe was akin to studying the mind of God, and once he wrote: "This most beautiful system of the sun, planets and comets, could only proceed from the counsel and dominion of an intelligent and powerful being."[43] American philosopher of religion and science Kelly James Clark has argued: "Newton's updated physics required God as a scientifically necessary hypothesis: for Newton, good physics was also good theology."[44] The presence of cultural bias and philosophical or religious predispositions does not negate science. Indeed, eliminating these things is an impossible requirement and many historians of science recognize it. Dick, in criticizing Basalla's book, argued: "if everyone followed Basalla's logic . . . the scientific enterprise would never have gotten off the ground."[45]

So we can reject the premise that C/SETI is a pseudoscience on account of its connections to religious thinking. But, while Basalla took his conclusions too far, he was not incorrect in noticing the impulse by CETI scientists to speculate on the existence of "superior" or "advanced" beings who, like a God, might come to save us. This phenomenon seems to manifest itself in thought about extraterrestrial life more generally, even in the present day. In late March 2022, as Russia intensified its war on Ukraine, Robert Hayes, a nuclear engineer and a popular science educator at North Carolina State University, posted a video on social media asking: "There have been a lot of people commenting that the aliens are going to save us, that spacecraft are going to stop World War III . . . Are there really that many people out there that believe that?"[46] In recent days, Hayes has responded to questions that his audience asked about nuclear bombs; such questions were receiving renewed interest as a result of the Russian invasion of Ukraine and a return to Cold War-style anxieties. Hayes was nevertheless puzzled by the strange insistence that aliens would somehow resolve the conflict. He was not the only one who noticed this unusual assumption. Several tabloid newspapers published headlines with claims like these: "UFO Believers Think Aliens Will Stop Nuclear War"; "Did Mystery UFO Wipe Out Russian Tanks?"[47] That impulse of hoping to be saved – of intervention, not divine but extraterrestrial – can be found in C/SETI and may in fact be one of its defining features.

There are dozens, if not hundreds, of examples of C/SETI researchers expressing hope that contact with extraterrestrial intelligence might promote peace on Earth. Former US diplomat and SETI researcher Michael Michaud once observed:

> Appeals for guidance or intervention from beyond the Earth have been part of many human religions. Often, those appeals were directed to the skies where gods were believed to dwell. Advanced extraterrestrials, far more omniscient and omnipotent than we are, could have many of the characteristics now attributed to the supernatural God of monotheistic religions."[48]

Jill Tarter argued that a successful detection in C/SETI would have great predictive implications for life on Earth, as it would give us evidence that technological civilizations do not necessarily destroy themselves: "If we detect a signal, even if there's no information – even if it's just a cosmic dial tone – we learn that it's possible for us to have a future – a long future."[49] Drake also frequently suggested that contact with extraterrestrial intelligence might be akin to other moments in Earth's history where massive scientific or cultural progress occurred through knowledge transfer between civilizations. He once said:

> I fully expect an alien civilization to bequeath to us vast libraries of useful information, to do with as we wish. This "Encyclopedia Galactica" will create the potential for improvements in our lives that we cannot predict. During the Renaissance, rediscovered ancient texts and new knowledge flooded medieval Europe with the light of thought, wonder, creativity, experimentation, and exploration of the natural world. Another, even more stirring Renaissance will be fueled by the wealth of alien scientific, technical, and sociological information that awaits us.[50]

Basalla might have called such statements religious. But I would invite you to view CETI from a new historical perspective. It is true that C/SETI scientists sometimes projected their hopes for Earth's future onto extraterrestrials, but this is largely due to the legacy of the Cold War, not religion. The Cold War fostered existential anxieties – brought about by the realization that for the first time in human history we were capable of completely destroying ourselves. But it also fostered utopian thinking; the Space Race and the huge investment by the United States and the Soviet Union in scientific research and development (R&D) promised an exciting new future, a future made possible by science and technology. Many of SETI's contemporary ideas about technological "supercivilizations" and extraterrestrial rescue from global annihilation developed during the Cold War period and persist to this day.

Like Crowe, my own approach to this subject certainly borrows from the history of ideas, but will develop a more strongly materialist perspective, which focuses on the infrastructures that influenced the development of CETI philosophies. In other words, rather than arguing that the philosophy influences the science, I argue that scientific and political infrastructures influenced the philosophies. Anthropologist Brian Larkin defines infrastructure as "material forms that allow for the possibility of exchange over space," a definition that suits Cold War CETI, as its infrastructures aimed to create exchange both over physical spaces on Earth and in cosmic space.[51] Furthermore, CETI developed around the same time in the United States and the Soviet Union, largely driven by the development of new technologies, especially radio astronomy, signals intelligence, and space-tracking infrastructure. With Larkin's interpretation of infrastructure in mind, I will reveal how radio astronomers and CETI scientists interacted with Cold War infrastructures, especially infrastructures of communication, including conferences, international correspondence via telegrams and the post office, and surreptitious communication through relationships with intelligence-gathering communities.

While the studies of Dick and Crowe discuss the development of CETI as part of the ongoing tradition of the extraterrestrial life debate, my study gives CETI full attention, firmly placing its development in the Cold War. In the chapters that follow, I intend to provide a comprehensive study of how CETI was influenced not only by the history of the extraterrestrial life debate but by Cold War anxieties, globalism, and the onset of military radio intelligence, as there is ample evidence that these were significant in its development.

Both of Drake's first experiences with radio CETI (that is, the accidental detection at Harvard and his deliberately planned Project Ozma) made detections of signals that were likely of military origin. The early 1960s, when Ozma was carried out, was also a period when the United States invested in launching satellites for military and intelligence-gathering purposes; just as the search for extraterrestrial artificial signals began, there was a large increase in artificial extraterrestrial signals – they just happened to originate on Earth.

The ~~Elephant~~ UFO in the Room

At this point in the chapter, you may be starting to wonder: how come UFOs haven't been mentioned yet? This is a book about aliens, right? Where are the descriptions of flying saucers, the Roswell crash, or Project Blue Book?[e] Surely this fits with the scope of the history of extraterrestrials and the Cold War. Was it not around the time of the Cold War that UFOs began to dominate the popular consciousness? There must be a correlation between the aforementioned fear of invasion by the "other" and the proliferation of UFO sightings. And golly, it sure looks like there

---

[e] Project Blue Book is the name of a US Air Force investigation into UFO sightings and related phenomena that was conducted from 1952 to 1969.

is a military connection with all of that. Not to mention the fact that even today NASA is dedicating resources to the study of unidentified aerial phenomena (UAPs)...

Ok, ok, you got me. It is a reasonable question. But you won't find any UFOs in this book. It is not because I think the UFO question is a historically uninteresting one (it is very interesting, and I recommend works by many of my talented colleagues, including David Halperin, Sarah Scoles, and Greg Eghigian).

The truth is, the histories of C/SETI and UFOs are inextricably intertwined. This is a fact that makes some people in the SETI community uncomfortable. C/SETI has long struggled to be taken seriously in the scientific community and its connection to the pseudoscientific, science fiction, and even the spiritual sometimes comes to the detriment of the small field. Thus some SETI scientists work hard to distinguish their work from ufology. But the boundary is often fuzzy at best, and C/SETI scientists have been involved in the UFO story since day one.

For example, in 1961, when he was just 27 years old, Carl Sagan was a surprise witness in the trial of Reinhold O. Schmidt, a UFO contactee and, later, a convicted fraudster. As part of a larger scheme to solicit money from the elderly, Schmidt claimed to have been contacted by Saturnians, who had selected him as their intermediary. As the scientific witness, Sagan argued that, due to differences in the atmospheres, surface gravity, and temperature between Saturn and Earth, it was highly unlikely that humanoid aliens like the Saturnians described by Schmidt could exist on the planet Saturn. Sagan's career was marked by discussions of UFOs from a young age.

While he was a prominent public skeptic, he was not immune from imaginative musings on alien visitations. In a book he co-authored in 1966, Sagan recounts episodes from Mesopotamian history that suggest encounters with non-human beings, stating that such historical encounters "deserve much more critical studies than have been performed heretofore, with the possibility of direct contact with an extraterrestrial civilization as one of many possible alternative interpretations."[52] Thus it is clear that, despite protestations from scientists, one cannot perfectly separate CETI from the UFO question.

Despite this historical connection, however, historians (myself included) generally make a distinction between C/SETI and ufology.

There are two good reasons for doing so. The first is that the C/SETI community grew specifically out of the discipline of astronomy, while UFO studies did not. The two communities do not generally overlap; therefore they merit distinct studies. Astronomy concerns itself with phenomena that occur outside our planet's atmosphere, while UFO studies focus largely on phenomena that happen inside our atmosphere. And the second reason why C/SETI and UFO studies should be considered distinct from each other is that the nature of the communities is different and the communities themselves wish to be considered distinct from each other. C/SETI is historically a close-knit community with a largely singular goal: to detect evidence of extraterrestrial intelligence using the tools and methods of science. Ufologists hold diverse beliefs and often lack connections with one another, spanning a wide spectrum of views, individuals, and groups.

An ad hoc committee advising the Permanent Committee on SETI for the International Academy of Astronautics produced a set of recommendations for addressing the UFO problem.[53] Ultimately they recommended that the SETI committee not concern itself with matters related to UFOs, on the basis that SETI "has historically focused on astronomical and social aspects of SETI, which use different methods (astronomy and social science)" from those "employed in UFO studies," and also "that UFO studies in general have not met the necessary evidentiary criteria for demonstrating that UFOs might have anything to do with alien technology." Ultimately, the contemporary SETI community strives to set itself apart from ufologists, with reasonable justification.

It is true that these boundaries are arguably getting fuzzier with time, especially considering NASA's recent involvement in examining the cause of unidentified aerial phenomena (and the fact that historically NASA has supported, albeit lukewarmly and sporadically, CETI but not ufology). There is a lot of potential academic merit to studying the UFO question. But my goal here is to create a history of Cold War CETI by focusing on a distinct cast of historical actors and by examining the field from a vantage point located within the astronomy community. So, no matter how tantalizing the issue of UFOs may be, let us set it aside for now.

A Critique Is Not (Necessarily) a Condemnation

So far, I have surveyed the various and fraught attempts by historians and scientists to characterize CETI. Some, like Basalla, have done so in a highly critical manner. In the chapters that follow, I too will critique CETI and its ideas. I will critique its practitioners who, as human beings, are inevitably flawed, sometimes hypocritical, sometimes biased, sometimes harmful. Unlike Basalla, however, I strive to do so not to delegitimize the field and its practitioners, but to better understand their motivations, successes, and mistakes, in the hope that understanding these characteristics might bring even greater value to the field. But critiquing CETI (and other sciences) can sometimes come with unpleasant consequences.

In 2021 I published a paper that examined the history of radio astronomy and C/SETI and its relationship to colonial heritage.[54] Specifically, it looked at how astronomers involved in C/SETI have historically used language and metaphors rooted in frontier myths and colonialism to frame their conceptions of "first contact" with extraterrestrial civilizations. Drawing from the history of orientalism and the US frontier, my paper explored the influence of cultural power structures such as gender and colonialism on C/SETI's physical and disciplinary foundations, highlighting how these factors shaped the scientists' pursuit of finding and communicating with alien intelligence. It concluded with a recommendation that present-day SETI researchers engage critically with the history of colonialism and warfare when formulating new search strategies and theories on intelligent civilizations, as doing so might assist them in removing anthropocentric and Eurocentric bias from their research.

Fortunately for me, the paper was popular within the academic community and the editors of *Scientific American* published an interview with me on the subject. The magazine article, like my original paper, argued that SETI should not focus only on looking outward for signs of cosmic civilizations, but also look inward to better understand how our own preconceptions can shape our perceptions of whatever we ultimately find.

A good number of people became quite angry. A few notable scientists even took to the internet and media to castigate me. A retired professor of biology wrote a rambling essay about me on his personal website, lamenting that science is becoming "wokeified" and protesting: "I don't think Charbonneau knows how SETI works."[f][55] One physicist published multiple angry articles online, one referring to me and others who conducted research in these areas as "lunatics" (but, as Billy Joel once wisely professed, "it just may be a lunatic you're looking for").[56] While it may be easy to ignore the hostile outbursts of those who are easily distressed by shifting cultural tides and by interdisciplinary discourse they do not fully understand, I was still troubled that a number of professionals seemed to take my critique of C/SETI as an attack on science.

Even more was I concerned that this work was viewed by some as an assault on the contemporary SETI community. This was an unhappy surprise for me, as I regard my work as supporting SETI. In fact I am an active member of the SETI community, a scientific staff member of an astronomical observatory, and have spent nearly a decade researching C/SETI. I have, in no uncertain terms, dedicated my life and career to researching and supporting science, especially radio astronomy and C/SETI.

The backlash came in part because the *Scientific American* article used the term "decolonization" (although my original paper does not) to describe some of the efforts to foster critical thinking on colonialism and science. The word "decolonization" has lost much of the context of its origins in academic and activist circles and is now seen as a sort of "boogeyman" term signaling racial conflict. And I can understand how, in the different context of C/SETI, it might initially sound rather silly – how can we "decolonize" a science that has seemingly nothing to do with statecraft or race, or even (in theory) with humanity? Furthermore, many people are unfamiliar with colonial history – it is not effectively taught in US schools, especially as some states are overhauling their history programs to focus on instilling "patriotic" values instead of promoting the development of analytical skills. So, given the lack of education in the United States on this history, as well as public sensitivity to discussions on racial topics, perhaps a volatile reaction to the interview should not have been surprising. But I *was* concerned to see how many people, especially professional scientists, did not seem to appreciate that providing tools to interrogate bias is a critical way to support science, not to tear it down.

---

[f] Reader, I promise that I do know how SETI works.

We need only look to the history of science to understand how much the development of tools to combat bias has aided science. In the eighteenth century, German physician Franz Mesmer had popularized animal magnetism, a healing technique that involved giving patients a high dose of iron and then running a magnetized object over their bodies. Mesmer was not an occultist – he believed his techniques and theories to be scientific. In 1766 he published a doctoral thesis titled "Dissertatio physico-medica de planetarum influxu" ("Physical–Medical Dissertation on the Influence of the Planets"), in which he argued that all living beings possessed a special "magnetic fluid" that was the foundation of life and was subject to the forces of gravity and magnets. By manipulating this fluid with his magnetic technique, he could "unblock" it and heal injuries and illness.

To the surprise of the medical establishment, Mesmer's techniques appeared to sometimes work – people's illnesses would improve; one woman's blindness was even temporarily healed. Mesmer's methods even appeared to put his patients into a trance-like, convulsive state (interestingly, this is where the word "mesmerized" originates). Mesmer's scientific contemporaries were rightly skeptical that human bias was affecting the outcomes of his healing, so in 1784 Louis XVI, the French king, established a royal commission headed by the American polymath Benjamin Franklin and the French chemist Antoine Lavoisier, to study Mesmer's technique. In the course of this endeavor they developed a new tool in support of the scientific method: the blind study (Figure 1.4).

Franklin's commission blind-folded some of the participants and exposed them to objects that either were or were not magnetized. Patients who were not being magnetized (but believed they were) still fell into convulsions, whereas patients who were being secretly magnetized without their knowledge experienced no effect. The commission's blind study not only discredited Mesmer's technique as a pseudoscience; it also helped discover what is now called the placebo effect – bias that is so powerful it can have a physical effect on our bodies.

* * *

We often consider the elimination of bias as more relevant to the social sciences or the "soft" sciences. Yet it also affects the so-called hard sciences such as mathematics and physics in surprising ways. Consider

Figure 1.4 This engraving, entitled "Le Magnétisme dévoilé" (1784–5, artist unknown) depicts Benjamin Franklin as the head of the royal commission, leading the fight against pseudoscience. Credit: Bibliothèque nationale de France, département Estampes et photographie, RESERVE FOL-QB-201 (115).

the Hubble constant. For decades, scientists have been refining their measurement of this fundamental parameter, which represents the rate at which the universe is expanding. The problem is that the standard cosmological model predicts a given value, but when cosmologists measure that value using observations in the local universe, they come up with numbers that contradict what the standard model predicts.

Cosmologists like the simplicity and beauty of the standard model of cosmology. When they test its predictions and find that the observations are discrepant with the standard model, some suspect that the observations are in error and the model is correct, while others believe that the

WHAT IS CETI?

model needs to be modified. Cosmologists tend to favor methods that support their expectations (don't we all?).

These scientists are getting creative and brave to find solutions to combat this well-known problem of confirmation bias. The H_0LiCOW[g] collaboration, led by astrophysicist Sherry Suyu, uses a unique method of calculation that is designed to avoid subconscious experimenter bias. The team developed an analysis technique that "blinds" the value of the Hubble constant as its members process their data, "opening the box" only when the analysis is complete. They agreed to publish the result, no matter the outcome. In other words, they hid the end value from themselves, as they worked to calculate it, in the hope that this would minimize biasing the results.[57]

What does this tell us about bias in science? Well, with the Hubble constant problem, scientists are simply trying to calculate a single number – nothing could be more quantitative, seemingly less subject to bias. And yet bias is still a huge problem. And, rather than ignoring that problem, they came up with creative solutions to minimize their subjectivity. In addition to the H0LiCOW collaboration's unique calculation method, cosmologists also limit bias by including as many different methods of observation as possible, including observations of cepheids, masers, and surface brightness fluctuation, to try to find the same answer. In short, they hope that being conscious of bias and expanding their scope will lead to more accurate scientific results.

And this is, in part, why SETI is benefited – not harmed – by attempts to rigorously examine bias in its history. SETI is one of the most creative and interdisciplinary scientific fields in existence; this is so by virtue of the fact that we are not entirely certain what we should be looking for. For all we know right now, the only intelligent civilization that uses radio telescopes is the human one. This has led SETI scientists to attempt to remove themselves from our "anthropocentric" mentality and imagine the variety of worlds and civilizations that might exist in the universe. It turns out, however, that removing oneself from one's cultural framework is more easily said than done. As you will see throughout this book, CETI scientists struggled with the inherent challenge of overcoming

---

[g] The Hubble Constant is represented by the variable H_0. The full title of the collaboration is "H_0 Lenses in COSMOGRAIL's Wellspring" – hence H_0LiCOW. (Astronomers are silly people.)

anthropocentrism and cultural biases in their pursuit of understanding and detecting extraterrestrial intelligence. And critically, although CETI was largely populated by researchers working in the physical sciences, they frequently engaged in social scientific or humanities problems, such as speculation on the nature of intelligence, civilization, and forms of contact in human history. CETI scientists often strove to offer a holistic perspective on the history of life in the universe and to integrate scientific knowledge and historical understanding to offer a comprehensive framework for exploring the history of the cosmos and humanity's place within it. In doing so, they often attempted to create a single and self-contained narrative about the universe and humanity, to find universal patterns and truths across great periods of time and geographical locations, and even to project these narratives onto extraterrestrial civilizations.

This approach has frustrated historians, for we have long been a part of the CETI community. In fact, at that US–USSR conference in 1971, a historian from the University of Chicago, William McNeill, was in attendance. According to Drake's recollections, McNeill "became extremely upset by the discussions" held by astronomers during the conference.[58] Drake credited this upset to McNeill's lack of ability to imagine the potential diversity of civilizations in the universe. McNeill remembered it differently. His participation in the conference was largely critical, and he later wrote in the *University of Chicago Magazine* an article, pointedly titled "Journey from Common Sense," in which he described his experience at the conference as largely bewildering.[59] McNeill had been invited with the expectation that he would synthesize Earth civilization and history with a cosmic projection. But the historian was frustrated by the superficial engagement the astronomers had with history, especially the way in which they extrapolated deterministic paths from the past and projected those not only onto our human future but onto the entire cosmos. Take, for example, my earlier quotation from Drake in which he speculates that contact with extraterrestrial intelligence might be comparable to the Renaissance. A historian may rebut that actually the transition from the Middle Ages to the Renaissance in Europe has been simplified and misunderstood, since a lot of new historical research proves the medieval period to be just as vibrant and exciting as the Renaissance. The historian might also point out that, for every example of a positive cultural exchange on Earth, there is at least one counterexample that was deeply tragic, where knowledge was lost, not created. And, of course, there is no historical analogue on Earth for contact between two intelligent technocratic *species*.

It all goes to show that this book is part of a long legacy and great tradition of historians annoying astronomers.[h] Instead of a "blind" approach to combating bias in science, which may well work for problems such as calculating the Hubble constant, CETI necessitates the opposite approach: taking a hard look at history, so that its practitioners may recognize cultural and historical bias when it arises in their work. Here the value of being a skeptic lies not in trying to eliminate the bias altogether but in being highly conscious of it, a method found more commonly in the practice of social, not physical, sciences. But the strong connection between history and CETI demonstrates that historical critique can function as a useful tool to combat bias and misconception in science. Carlo Ginzburg, an Italian historian, helped popularize a subfield of history known as microhistory, a sort of antithesis to the "big history" promulgated by CETI. Microhistory is a historical approach that focuses on studying and understanding specific individuals, events, or small communities within a larger historical context. Instead of emphasizing broad trends and overarching narratives, microhistory zooms in on the details, offering a deep analysis of specific case studies. It seeks to uncover the intricacies of social, cultural, and individual experiences, shedding light on the complexities and nuances of the past. It rejects "universal" narratives and generalizations, in order to poke holes in our assumptions about history. Ginzburg once argued:

> The historian's task is just the opposite of what most of us were taught to believe. He must destroy our false sense of proximity to people of the past because they come from societies very different from our own. The more we discover about these people's mental universes, the more we should be shocked by the cultural differences that separates us from them."[60]

In this spirit, the historian's job is very much like that of the C/SETI scientist. We aim to find the alien in the familiar, to question the expectations and preconceived notions we hold about our world, to explore

---

[h] My husband, the astronomer married to an annoying historian, would heartily agree.

strange *old* worlds and civilizations. If this were to be a microhistory (and it is decidedly not), it would be a microhistory focused on that singular conversation between I. S. Shklovsky, Carl Sagan, and Frank Drake in 1971. What can we learn from that particular moment in time, at the start of détente in the middle of the Cold War, a conversation between Soviets and Americans on the nature of communication across vast gulfs – the Iron Curtain, or the curtain of outer space?

2

Alien Intelligence
Radio Astronomy and the Dawn of the Cold War

The astronomer has detected a signal from an alien intelligence. She is observing with the Very Large Array (VLA) at the National Radio Astronomy Observatory (NRAO), when she notices a series of prime numbers embedded in the signal she was receiving from the distant star system of Vega, 25 light years away. Upon further investigation, she discovers that there is more to these numbers: there is a video hidden in the signal! She and her team gather around a screen to watch, shivering in anticipation. Are they about to view the first ever message from an intelligent extraterrestrial civilization?

A blurry symbol appears on the screen. After zooming in and out to adjust the image contrast, they are able to bring the symbol into focus: a swastika on the arm of Adolph Hitler. The video plays, showing Hitler's opening address at the 1936 Summer Olympics in Berlin, Germany. The astronomers are shocked. Is it a hoax? The alien signal was a message from us, our history, all along.

This scene, from the 1997 film *Contact*, written by Carl Sagan, highlights themes that recur in CETI's history: a connection between science and warfare; finding artificial signals that originate from Earth in our pursuit of extraterrestrial ones; confronting our own history in our attempts to contact extraterrestrial intelligence.

In the film, the astronomers eventually discover that the signal they detected was in fact from an extraterrestrial civilization; the aliens had just been using the Hitler clip to grab our attention. The 1930s may have marked the first decade in Earth's history when we became radio "noisy": television signals at that point in history might have been strong enough to pierce Earth's ionosphere and carry on into space. It is possible, in Sagan's view, that the address by Hitler was one of the first artificial radio signals that emanated from Earth into the cosmos. What a poor first impression, if that was the case.

But Sagan was shrewd in highlighting a truth about CETI: that it is difficult to distinguish between artificial signals from Earth and those we might find coming from extraterrestrial intelligence. This was especially true at the time CETI developed. While it could have been the case that since the 1930s humans developed artificial radio signals strong enough to leave Earth's atmosphere, it was not until the Cold War period that we began to put actual antennas into space, in orbit around our planet.

What linked Cold War military activities with the search for and communication with extraterrestrial intelligence was a desire to identify a strong distinction between artificial and natural signals. Artificiality is a key component of intelligent signals. The artificial is distinct from the natural in that it carries intrinsic information and is constructed deliberately and with intention, rather than occurring organically or spontaneously. The shared motivation to seek out artificial signals created an incentive for cooperation; instead of there being a distinction between the pursuits of CETI and those of the military, the means were sometimes one and the same. Radio astronomy and CETI benefited from the military, and the military benefited from CETI and radio astronomy; these sciences were inextricably tied up in the military–industrial complex – and particularly in intelligence gathering. Often the same astronomers who cooperated with their international peers also constructed the tools they used for spying on their colleagues.[1] Watching and being watched were intrinsic parts of collaborative scientific work during the Cold War, and CETI is a particularly revealing example of this fusion of scientific and geopolitical ambitions.

But the relationship between CETI and the military was a contentious one. Radio astronomy and CETI as we know them would not exist without the Cold War; as we shall see, the Cold War provided funding, technological innovation, political incentive, and philosophical motivation. Yet it would be too simple and bombastic to argue that radio astronomers and CETI scientists were military sciences. The reality is far more nuanced, and many radio astronomers and CETI scientists would rightfully protest being labelled military scientists. The presence of artificial signals in space from the military often frustrated CETI scientists; as we saw in Frank Drake's early CETI attempts, secret military signals could make for accidental false alarms in the pursuit of artificial signals from extraterrestrial intelligence.

It was certainly true that much of the infrastructure supporting radio astronomy and CETI either was military in origin or was used, at least in part, to support military activities. But while some radio astronomers and CETI scientists willingly cooperated with the military, others were outraged at military interference in what they viewed as an apolitical (and even peace-making) scientific pursuit. Some astronomers, on the other hand, had no choice in their military involvement; in the case of the Soviet Union especially, there was no flimsy distinction between civilian and military facilities, as there was in the United States. Soviet scientists who wished to cooperate with their international peers and travel abroad were often obligated to adhere to political and militaristic agendas and were frequently compelled to work on military projects alongside their science. The Cold War environment was tense and fraught; this tension is what promoted the development of tools and funds in support of sciences like radio astronomy, but it also made pursuing science difficult.

Falling Behind

The first US government observatory for radio astronomy, NRAO, was founded on November 17, 1956.[2] The establishment of NRAO in Green Bank, West Virginia was in part a response to the newfound scientific–technical competition between the United States and the Soviet Union – prompted by a growing sense that the United States was falling behind in the field of radio astronomy, which as we shall see had greater implications for its warfare capabilities at the start of the Cold War. As noted by historian Paul Forman, US physics in the 1950s "underwent a qualitative change in its purposes and character," as a result of increased government intervention and a new emphasis on "the nation's pursuit of security through ever more advanced military technologies."[3] In turn, the government's intervention in the sciences was driven by the growing military threat from the Soviet Union, which, like the United States, was expanding its nuclear capabilities. But the investment in radio astronomy was also tied to earlier military activities during World War II, and in particular to the development of radar.

The first operational radar system was developed by Robert Watson-Watt from 1935 to 1938, as the United Kingdom was on the cusp of war with Germany. Watt's work was supported by the British Air Ministry because of the government's recognition that Britain was vulnerable to the growing threat of an air attack by the German Luftwaffe.[4] Radar worked by using a transmitter to bounce radio waves off distant objects to receive accurate information on said object's position and speed, even when that object was not optically visible – for example at night, when most airstrikes occurred. As the war expanded across the world, most of the countries involved (especially the United Kingdom, Germany, the Netherlands, Australia, the United States, the Soviet Union, and Japan) had developed radar systems. These nations recruited physicists and radio engineers, often from universities, in an early case of what is popularly known as "the military–industrial complex," a term coined by President Eisenhower in his final public speech in 1961.[5] Since many of these military systems developers were trained as academic scientists and engineers, they sometimes made serendipitous scientific discoveries related to their military activities. One prominent example would be the observations of James Stanley Hey, a British physicist who joined the British Army Operational Research Group (AORG), which aimed to improve the capabilities of anti-aircraft radar systems.

Because of the efficacy of radar as a defense system, it should be noted that, during the war, countries often tried to "jam" their enemy's radar systems – in other words to send out radio signals that interfered with those systems by flooding them with noise or other false information. The AORG studied both how to conduct and how to fend off radar jamming. On two days in February 1942, it discovered that ten radar systems were being jammed through overwhelming noise interference in the 55–85 MHz range, which, Hey observed, "came almost continuously and exclusively between dawn and sunset."[6] Upon further investigation, Hey noted:

> Although no explanation of how the interference was caused can be given, it does not appear possible to arrive at any other conclusion ... than that the interference was associated with the sun and the recent occurrence of sun spots.[7]

As the war came to a close, Hey published his findings about the radio emissions from the Sun in the scientific journal *Nature*. This was one of the earliest cases of military radar operators transitioning into radio astronomy research in the postwar period.[8]

Another notable example of a wartime radio physicist transitioning into radio astronomy research at the end of the war is Sir Bernard Lovell. Lovell was a British physicist working in cosmic ray research at the University of Manchester. At the start of the war in 1939, he joined the Telecommunication Research Establishment (TRE) to conduct research into H2S radar systems.[9] While working on one of these radar systems, Lovell noticed background "echoes" when trying to observe aircraft signals, and hypothesized that the echoes could be caused by cosmic ray "showers."[10] At the end of the war in 1945, he returned to Manchester and used ex-military radar equipment to continue his studies of cosmic rays, in the hope of identifying the source of the echoes. But instead of finding cosmic rays Lovell discovered that the radio signals were caused by meteors, which ionized small parts of the atmosphere during their descent.[11] This observation made in 1945 was the start of one of the world's earliest and most distinguished radio astronomy observatories: Britain's Jodrell Bank Observatory.

Historian of science Jon Agar has written extensively on the establishment of Jodrell Bank and argued that the development of radio astronomy in Britain benefited from the trend in nuclear physics to establish centralized research facilities that were disconnected from traditional university settings. On the other hand, the United States' own centralized radio astronomy facility, NRAO, would not be established until over a decade later, despite this country's prominent role in the war. While Agar noted that radio astronomy in Britain grew "disproportionately quickly" as a result of its wartime activities, shifting scientific organizational structures, and interest in prestigious symbols of scientific–national power, similar arguments could certainly be made for the United States. Nevertheless, the United States lagged slightly behind Britain in its interest in radio astronomy. Why?[12]

NASA senior economic advisor and historian of astronomy Alexander MacDonald wrote a history of what he called "the long space age" that charted the enormous financial investment in optical astronomical observatories in the United States during the nineteenth century. There he argued that "the driving motivation for the provision of funds was a desire to signal status and capability through monumental achievement."[13] There was, then, in the United States, a strong connection between prestigious optical astronomical observatories and elite status. Historian of radio astronomy Woodruff Sullivan has argued that the United States' eminence in optical astronomy "acted as a deterrent to the fledgling specialty" of radio astronomy.[14] Two key reasons contributed to this. First, the optical astronomy community received little federal funding during the war and in the early postwar period, especially by comparison with other areas of physics. Secondly, optical astronomers did not view radio astronomers as "real astronomers" because of their different disciplinary backgrounds: at that time, radio astronomers tended to begin their careers as radio engineers or physicists, and these were not traditional astronomy backgrounds. In a view from across the pond, British radio astronomer and later Nobel prize winner Martin Ryle observed:

> The gulf between radio and astronomy was probably worse in America. . . . I think it's probably fair to say that the [American] optical astronomers weren't particularly welcoming. There was a bit of a closed shop, [the radio people] weren't real astronomers. They hadn't been brought up properly at all . . . they didn't know what Oort's constant was, they didn't know Hubble's constant.[15]

Furthermore, many US astronomers believed that radio astronomy would be an empty and meaningless pursuit; they did not believe that there was anything interesting to observe in the radio and that much more was to be gained by studying the optical. In fact, even NRAO's first director, the esteemed optical astronomer Otto Struve, believed that the universe was largely radio-quiet and warned that radio astronomy was probably "barren" in comparison to optical astronomy and would eventually fail as a discipline.[16]

A third reason for the gap between US and international radio astronomy development stemmed from national institutional frameworks. In countries like Britain, radio research in the prewar era was situated mainly in physics departments. This made it easier for researchers to transition to solar studies and eventually to radio astronomy upon return to their universities after serving in the war effort. In the United States, on the other hand, radio and ionospheric research was usually situated either in electrical engineering departments at universities or in specialized institutes such as the Naval Research Laboratory or Bell Labs, which struggled to establish strength in radio astronomy – in part because of their emphasis on practical research, in part because of their military ties.[17] This gap between radio astronomers and optical astronomers in the United States led to a decade-long lapse in radio astronomy research, while researchers in other countries, especially Britain, Australia, and the Soviet Union, blazed ahead.

Before moving on to examine the urgent and rapid expansion of US radio astronomy in the late 1950s, it should be noted that the Soviet Union's early accomplishments in this field were a different case from both that of Britain and that of the United States. While it may have been true that early radio astronomy technology in the United States and Europe was of military origin, the transition from military use to scientific use first occurred largely in universities, when wartime radio operators returned to their civilian lives, bringing formerly military equipment with them. In the Soviet Union, on the other hand, the tools and equipment remained largely in military control. This had major implications for the ability of Soviet radio astronomy results to reach a global audience. And because there was no transition of radar equipment to civilian and scholarly facilities, initially the Soviet Union became much stronger in the theoretical than in the observational study of radio astronomy.

The Soviet emphasis on theory was a feature in both optical and radio astronomy, which perhaps explains why there was less resistance to integrating radio astronomical research at major institutions in the Soviet Union such as Pulkovo Observatory, Sternberg Astronomical Institute, and Lebedev Physical Institute. Early and prominent radio astronomy theorists such as I. S. Shklovsky began publishing on radio astronomy topics shortly after the end of the war. From the perspective of the United States, the Soviet Union was largely a closed and mysterious society at this point. Because of the presence of what Winston Churchill, the British prime minister, famously baptized "the Iron Curtain," a sense of anxiety began to grow in the United States that the Soviet Union was making great advances in radio astronomy.

This begs a question: why did the United States care? As already noted, the United States dominated research in other areas of astronomy and was already establishing its role as a leading scientific superpower. Furthermore, there were some major early accomplishments in US radio astronomy: the US radio engineer Karl Jansky is credited with making the first deliberate observation of cosmic radio waves, and the US radio enthusiast Grote Reber is credited with building the first parabolic radio telescope.[18] Why would there be so much anxiety about success in this newfound and previously disregarded subdiscipline? The scientific competition narrative, which has already been well studied by historians of science and the Cold War, is one answer.[19] As part of its campaign to achieve global dominance and to signal to "unaligned" countries its technological superiority over the Soviet Union, the United States invested heavily in the sciences at the start of the Cold War. Scientists were not hapless bystanders; many recognized the political atmosphere as being helpful to their goals.[20] It is therefore possible that some astronomers exaggerated the deficiency of radio astronomy in the United States and used the "falling behind" narrative to extract more resources from the government.

But the start of the Cold War, the establishment of Jodrell Bank, and the launch of Sputnik in 1957 also demonstrated a dual use of radio astronomy facilities: in addition to their scientific capabilities, they could also gather intelligence, a technique that would become a defining feature of the "cold" warfare that was to span the subsequent decades. Thus radio astronomy in the late 1950s transitioned from being a quirky postwar science experiment with discarded military equipment to gaining the status of a titan in government-funded science. The interconnected nature of radio astronomy and the military would have major implications for scientific communication and the development of CETI in both the United States and the Soviet Union.

Country Roads[a]

If any city might be considered the birthplace of CETI, it would no doubt be Green Bank, West Virginia. The word "city," though, is probably generous; the 2020 census lists a population of 141 residents there.

---

[a] A reference to John Denver's hit song "Take Me Home, Country Roads" (1971), which is an homage to the state of West Virginia.

Figure 2.1 Green Bank, West Virginia. Credit: Rebecca Charbonneau.

Green Bank is a small rural town in the heart of Appalachia, marked by anachronisms and contradictions. The town is located in a highly remote area, surrounded by hundreds of miles of mountains and dirt roads. But it is also home to one of the most sophisticated scientific and technological endeavors in the United States. NRAO was founded in Green Bank in 1956. Today, the site is occupied by dozens of telescopes, including the Green Bank Telescope, which has a 100-meter diameter, is nearly 500 feet tall, and provides a view into the radio universe (Figure 2.1). The town is a window into a rustic past, but also into a cosmic future.

Green Bank is considered the birthplace of CETI for number of reasons. It is where the first scientific CETI project, Project Ozma, took place. One year after Project Ozma was conducted, in 1961, the first ever conference on communication with extraterrestrial intelligence took place, also in Green Bank. From 1960 until today, Green Bank holds the title of conducting more searches for extraterrestrial intelligence than any other observatory.[21] The Observatory[b] is proud of this fact – if you were

---

[b] A small detail that needs pointing out: in 1966, ten years after NRAO's foundation, its headquarters moved from Green Bank to the town of Charlottesville in Virginia. The telescope operations remained in Green Bank, but Charlottesville was a more suitable place for the headquarters partly on account of its proximity to Washington DC, where NRAO's funding bodies were located. Charlottesville was also an easier place for the scientists to live in, being much more metropolitan than Green Bank (although still small and fairly rural). The telescopes and facilities at Green

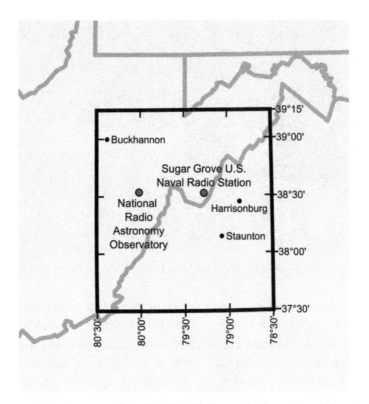

Figure 2.2 This is a map of the NRQZ, with the marked location of the original headquarters of NRAO and Sugar Grove. The NRQZ spans approximately 13,000 square miles near the state border between Virginia and West Virginia. Credit: Wikimedia Commons.

to walk around the main offices in Green Bank, the Jansky Laboratory, you would see posters that read "The Universe Is Whispering to Us" and "Life Is Possible beyond Our Planet." If you are lucky enough to be allowed to climb to the top of the Green Bank Telescope, you might peer down through the grated walkways and spot a little green figure with a big head and bulging black eyes. John Denver once sang, in his renowned homage to the great state of West Virginia, "the radio reminds me of my

---

Bank continued to be operated by NRAO until 2016, when the National Science Foundation modified the contract for the site and turned it into an independent observatory – the Green Bank Observatory (GBO). GBO still maintains a close relationship with NRAO. Because this book spans the period of time in which the site at Green Bank was operated by NRAO, "Green Bank" and "NRAO" will often be used interchangeably.

home far away." So too do Green Bank's radio telescopes remind us of the existence of other worlds far away; are they also someone's home?

But Green Bank's history concerns more than just CETI. Just as radio astronomy grew out of wartime activities and interest in World War II, the rest of its history, too, became mired in the military's. The Observatory at Green Bank was established with scientific goals in mind – it was not a military operations center. But the ghost of the Cold War has haunted Green Bank from its very beginnings.

The haunting began in 1958, when a silence fell over the border between Virginia and West Virginia. The Federal Communications Commission (FCC) had established an area of about 13,000 square miles as the National Radio Quiet Zone (NRQZ) (Figure 2.2). The NRQZ placed restrictions on radio broadcasting within its perimeter; any existing broadcasting facilities had to operate at reduced power and with highly directional antennas, to minimize overall radio noise.[22]

The principal reason given for this large-scale federal intervention in the electromagnetic spectrum was the establishment of the NRAO in Green Bank. The choice to site NRAO in Green Bank was a technical one; astronomers have long sited their telescopes in remote locations to produce the best observational results. In optical astronomy observatories, light pollution is a primary consideration in choosing an observatory site. Optical telescopes perform best when set in a dark, dry, and high-altitude environment, which minimizes interference from city lights, humidity, and atmospheric disturbance. Therefore many optical telescopes are located in secluded areas, away from large cities and their bright lights.

Although city lights are not much of a factor in the successful performance of radio telescopes, they too benefit from being located far away from large cities. This is because radio telescopes cannot perform meaningful observations if there is too much radio frequency interference (RFI). RFI is caused by sources that generate changing electrical currents that are detected by radio telescopes. Such disturbances may appear simply as elevated "noise" or as a strong local signal that overshadows fainter cosmic sources (a reminder here that this "noise" cannot be heard: radio astronomy deals in invisible light, not sound, so RFI simply makes it more difficult to identify cosmic signals in astronomer data). RFI can be caused by both human-made and natural sources, such as cellular networks, lightning, vehicle ignition systems, and radio towers.

Figure 2.3 RFI restrictions. Signs such as these are found around the premises of the Observatory in Green Bank, reminding scientists, visitors, and locals that technology that generates RFI is not permitted in the Radio Astronomy Instrument Zone, the most restrictive part of the NRQZ.
Credit: Rebecca Charbonneau.

Green Bank was a good choice of a site for NRAO, not only because its overall isolation decreased its exposure to RFI but because it was situated in a valley in the Allegheny mountain range, which provided natural shielding from such interference in nearby cities. Other steps, including the creation of the NRQZ, have been taken to further "quiet" the observatory. If you walk around Green Bank, you will see retro diesel cars roaming the town even today (they are more radio-quiet than newer

Figure 2.4 RFI mitigation truck. This is a picture, taken in 1981, of an RFI monitoring truck, which would drive around Green Bank and the surrounding area to identity sources of RFI. Credit: NRAO/AUI/NSF.

cars). You will see signs that warn against the use of mobile phones, personal cameras, and other radio-noisy devices (Figure 2.3). There is no wi-fi in Green Bank; there are even "wi-fi police" – RFI technicians who drive around looking out for unauthorized wi-fi use (Figure 2.4). If you were to decide that you would like a frozen burrito for lunch when visiting the Jansky Lab, you may be surprised to find the kitchen microwave hidden inside a large metal box – a faraday cage – to prevent the microwaves that heat up your burrito from escaping and wreaking havoc on the astrophysical observations being conducted that day (Figure 2.5). Critically, if you find yourself craving some groovy tunes, you may be

Figure 2.5 Microwave faraday cage. This is a microwave used by the Observatory staff in their lunchroom. The microwave is locked inside a special RFI-blocking box to prevent the microwaves from escaping and causing disruption to the nearby radio telescopes. The image is grainy and poor in quality because digital cameras are not allowed on the premises either: they, too, cause RFI. Hence a disposable film camera was used. Credit: Rebecca Charbonneau.

disappointed by the small selection on the radio; the only broadcast radio stations that exist near Green Bank are part of the Allegheny Mountain Radio Network, which operate stations using reduced power and highly directional antennas, in line with the requirements of the NRQZ. Life in Green Bank is unusual by comparison with life in the rest of the country, but the NRQZ is necessary to facilitating successful operations at the Observatory.

But the NRQZ did not serve an exclusively scientific support function. In 1958, buried in the last sentence of the fourth line item of the FCC's Docket No. 11745 – which amended the commission's rules and regulations to give protection from interference to frequencies used for radio astronomy – was this statement: "additional coordination would be undertaken by the commission with the Department of Navy at Washington, DC with respect to the Sugar Grove facility."[23] Although the FCC document gives no further detail on the purpose of the Sugar

Grove facility, and the National Reconnaissance Office still keeps many of the documents from the planning and development of facilities in Sugar Grove classified, it is today relatively well known from other sources that, soon after the establishment of NRAO, the Naval Research Lab began plans to build a 600-foot radio telescope for the purposes of gathering intelligence from the Soviet Union.[24]

Therefore, while the United States publicly promoted the scientific support function of the NRQZ, there was an undercurrent of military motivation. The brand of intelligence that the new facility at Sugar Grove was interested in conducting was known as "signals intelligence." Signals intelligence (SIGINT) is the branch of intelligence gathering that deals with the interception, analysis, and exploitation of foreign communications through the medium of radio emissions. The US Marine Corps manual defines SIGINT as "intelligence gained by exploiting an adversary's use of the electromagnetic spectrum with the aim of gaining undetected first hand intelligence on the adversary's intentions, dispositions, capabilities, and limitations."[25] In the Cold War period, SIGINT was used to target both seemingly benign sources of information, designed to fill in gaps related to lifestyles and activities in the mysterious Soviet Union, and sources of the country's military activities (such as ballistic missile testing); the latter aimed to gain an advantage in understanding (and therefore combating) Soviet weapons and war plans. Yet the trouble with SIGINT was that it was difficult to ascertain meaning in the absence of context or to know whether a given source of information was itself reliable. In an article on the history of signals intelligence and its applications, a CIA officer was quoted as having said: "Electronic intercepts are great, but you don't know if you've got two idiots talking on the phone."[26] In other words, the success of SIGINT depended on the ability to understand the subtleties of communication so as to determine the veracity, reliability, and correct meaning of the signal.

Despite this obvious potential for misinterpretation and miscommunication, a tremendous amount of governmental funds were sunk into SIGINT during the Cold War period. The authors of the aforementioned paper estimated that

> [the National Security Agency] and its predecessors ... spent about $100 billion since 1945, 75 per cent of which was spent on SIGINT and the rest on

communications security. More importantly, throughout the Cold War the US government spent four to five times as much money on SIGINT than [sic] they did on [traditional human intelligence] collection.[27]

On the Soviet side, it is estimated that SIGINT absorbed approximately 25 percent of the KGB's annual budget.[28] The Cold War marked a significant shift from traditional warfare's use of human intelligence gathering to intelligence that depended primarily on technology and the interpretation of technological signals. Just as with CETI, there was a newfound interest in searching for signals from foreign civilizations.

Sugar Grove is a particularly strange case in the history of signals intelligence, in part because of the transitory character of the period in which the station was established. In the late 1950s, as the facility was being designed and constructed, the plan was to collect intelligence from the Soviet Union by searching for Soviet signals that were reflected back to Earth from the surface of the Moon. The idea of using the Moon to capture artificial signals was not an entirely new one. In another case of military radio technology transitioning to scientific use, at the end of World War II John 'Jack' Dewitt, the director of the Camp Evans Signal Laboratory in the US Army Signal Corps, convinced the army to use its radar to conduct an experiment. The experiment, named Project Diana, aimed to use military radar equipment to send a radio signal to the Moon. The signal would then bounce off it and the team would measure its return, potentially paving the way for new modes of communication. In January 1946 the team succeeded. As one article put it, "the Project Diana team" had become "the first to send a signal into space and receive an answer back."[29]

Over a decade later, the Sugar Grove facility hoped to conduct a similar experiment. But this time it planned, as part of its SIGINT program, to build a 600-foot steerable telescope (the largest in existence at the time) to monitor Soviet radio signals that were unintentionally reflected off the surface of the Moon. It is important to note here that several astronomers, and in particular NRAO's former chief scientist Kenneth Kellermann, have expressed doubt that capturing Soviet Moon signals was the true objective of the Sugar Grove project.[30] It seemed strange that the US government would invest so much energy and money into a project of this scale that would be functional only for a fraction of the day, when the Moon was visible in both the Soviet and American skies at the same time. These suspicions may be correct – there are certainly examples from the history of the Cold War in which false stories were planted to obscure true ones. Yet I do not share the skepticism about Sugar Grove. As we shall see, there were many examples of "absurd," or even pseudoscientific, scientific projects funded or proposed by various governments during this period. The atmosphere of intense anxiety and paranoia during the Cold War would lead to several scientific projects with rationales that were unclear at least to some in the scientific community – and sometimes to a considerable number of its members. Historian of science Audra Wolfe has charted how, during the Cold War, science and technology were often seen as offering "the best solutions to the nation's problems, whether those problems might involve infrastructure or foreign policy."[31] This resulted in hundreds of millions of dollars spent on high-tech interventions that, from a contemporary perspective, seem now ridiculous. In her book *Competing with Soviets*, Wolfe points to physicist Edward Teller's proposal to use nuclear explosives as a "convenient" tool "for mining operations, oil and gas exploration, and . . . earthmoving projects" – suggestions that Wolfe argues are "a symbol of everything that was wrong with science and technology in the Cold War."[32]

Sugar Grove is another case of Cold War science gone wrong. The Naval Research Laboratory's plans for the 600-foot telescope never materialized. It was discovered that, as a result of poor planning and engineering, the base of the telescope was insufficient to support the enormous parabolic dish. Many parts of the dish would have to be replaced and made out of different materials, and a part of the telescope was to be redesigned altogether. Congress had already spent tens of millions of dollars on the project, but its redesign and reconstruction would cost even more; it would also lead to large delays in the eventual operation of the site. Finally, in 1962, the project was cancelled.[33] By 1962, the United States was becoming successful at regularly launching artificial satellites, some of which could be used for reconnaissance purposes. A satellite would be able to collect signals from the Soviet Union from directly above the country in low Earth orbit – so there would be no need to wait for brief windows of time when the Moon was overhead in both countries. The Naval Research Laboratory nevertheless continued to operate out of Sugar Grove, and the site was later purchased by the National Security Agency (NSA), to be used as a listening post.[34]

Clearly the establishment and maintenance of the NRQZ had a military motivation related to its use for intelligence gathering, even though it was largely promoted as an action designed to boost the success of civilian science. Although only the facility at Sugar Grove was explicitly military, both the development of the NRAO and the Sugar Grove facility should be understood as contributing to US geopolitical goals. This is because, while NRAO was (and still is) a civilian organization, its foundation was part of the larger federal investment in science and in the creation of scientific institutions in response to the Cold War. Additionally, while NRAO's main goal was conducting astronomical research, many of the tools and techniques developed by NRAO would have military applications, and vice versa; military facilities sometimes benefited the astronomers.

For example, we explored in the last chapter one of the earliest observations conducted at NRAO: Frank Drake's Project Ozma, which used NRAO's 85-1 telescope to search two nearby star systems for signs of extraterrestrial intelligence. The advent of radio astronomy and CETI science during the early Cold War is particularly interesting thanks to the use of its technology for spying both on the unknown intelligence of the cosmos and on the unknown intelligence here on Earth. Shortly after Drake arrived at NRAO and around the time when he was planning and executing Project Ozma, he and the other scientists at NRAO were invited to visit the Naval Communication Station in Sugar Grove, which was still actively building the planned 600-foot telescope.[35] Later on, in recollecting the experience in an editorial for the SETI League publication, Drake highlighted the mystery surrounding Sugar Grove, jokingly calling it "a major radio observatory almost no one has ever heard of," run by the Navy even though, as Drake cheekily put it, "where is the ocean?"[36] After all, Sugar Grove is hundreds of miles inland, nestled in the Appalachian mountains. Drake did not describe the exact nature of the visit or explain why NRAO scientists had been invited, but he described seeing "a bee hive of little rooms, all underground, protected from RFI by overhead imported charcoal and soil, with each room having a linguist with headphones on listening to conversations in faraway Russian."[37] It is not difficult to speculate on why NRAO scientists might have been invited to visit the facilities. The telescope project was underway and already facing challenges; it is possible that NRAO's new radio astronomers were asked for their expertise in telescope design or in some other aspect of collecting radio signals.

Drake's invitation to (and fascination with) Sugar Grove unveiled a growing link between the early days of military intelligence gathering and CETI. Drake wrote about Sugar Grove to the SETI League specifically because his visit many years earlier had sparked an idea. CETI scientists were interested in gaining an understanding of what Earth's technosignature might look like, for example its "radio leakage" – the signals from television, satellites, warfare, and any sort of radio emission activity that might be detectable to alien radio astronomers who pointed their telescope at our planet from their own. Understanding Earth's technosignature could aid CETI scientists in figuring out what "non-targeted" signals from other planets might look like (in other words, what to look for if not a direct and deliberately constructed message from a world other than our own) and in understanding how "noisy" our Earth might seem to other civilizations, which could be looking to say hello. If the facility at Sugar Grove was interested in "Moon bounce" signals, its scientists would have had to have an understanding of the radio noise atmosphere of the planet; and so, Drake argued, "there must be a huge collection of well-calibrated data on the radio signature of Earth as seen from interstellar distances."[38] Therefore, he speculated, "maybe there is no need to design space missions to gather data about Earth's electromagnetic leakage. The data already exists somewhere in Sugar Grove."[39] Drake's comment, though speculative, demonstrates the overlap between the aims of CETI and SIGINT – both striving during the Cold War period to detect artificial signals in space and, in doing so, developing tools and techniques that benefited their communities. He was not the only person to find the 600-foot telescope useful for CETI; the first scientific CETI paper, Morrison and Cocconi's "Searching for Interstellar Communication," proposed the use of the 600-foot telescope for the first CETI observations, given that it was designed for the purpose of seeking artificial extraterrestrial signals.[40]

Death by Telescope

After the war, Sir Bernard Lovell's fascination with using military radar to make atmospheric and astronomical observations led to a desire to build better equipment; after all, the ex-military tools he had been using had not been designed with exclusively scientific pursuits in mind. He initially had a 66-meter parabolic dish built out of wire that Robert Hanbury Brown, another British radar technician turned scientist, would use to discover radio waves that emanated from the Andromeda Galaxy.[41] But because the 66-meter telescope's dish was fixed (meaning that it could only face upwards and was not steerable), it relied on the rotation of the Earth to explore different parts of the sky and was therefore limited in the observations it could conduct. Lovell next began planning the construction of a 76-meter steerable dish, funded by the University of Manchester and the UK government. Unfortunately, Lovell's ambition pushed the initially £259,000 telescope budget to run approximately £381,000 over budget, leaving the university and government to cover the costs.[42] Shortly before the completion of the telescope in 1956, this project mismanagement caused Lovell to face an inquiry from the Public Accounts Committee that might have resulted in his imprisonment.[43] Fortunately for Lovell, however, in the year after his inquiry the Soviet Union launched Sputnik, the world's first artificial satellite.[44]

Because of the Soviet Union's status as a closed nation, there was initially great skepticism from the United States and European countries about Soviet claims to technological accomplishments – a recurrent theme in Cold War sciences. Sputnik itself could be tracked by conventional radio receivers and even by amateur radio enthusiasts keen on hearing its signature "beep-beep," but these tools were unable to successfully track the staged rocket that launched the satellite. Tracking the rocket was of the utmost importance to the western world – after all, a rocket capable of launching objects into orbit is not much different from an intercontinental ballistic missile. At the start of the Cold War's weapons race, it was crucial to state security to have rocket tracking capabilities.

To make matters even more difficult, Lovell was informed by the British Ministry of Supply that there was "no other radar facility in the West able to track the satellite's upper stage rocket," meaning that Jodrell

Bank was the only facility that would be capable of monitoring this new Soviet technological feat.[45] On orders from the British government, Lovell's newly built telescope tracked the Soviet rocket as it began its slow descent into the atmosphere, helping western nations understand what the Soviets had accomplished. The Mark I telescope was featured in headlines all over the world, and Jodrell Bank became an instant point of pride for the British. This event marked the start of what would become an important source of Cold War intelligence: space surveillance.

But, alas, the Earth rotates. Because of this, as Lovell tracked the Sputnik rocket, he reported to the *New York Times* that there would be moments when it would "not be visible" in the British skies, and therefore asked the United States to "take up the watch."[46] His request revealed an important theme in the history of astronomy: because of the nature of the Earth's rotation, continuous observation of an extraterrestrial source with ground-based telescopes is impossible without international cooperation. This fact will become important in understanding the unique internationalism of radio astronomy. But in 1957 few facilities were able to successfully conduct radio telemetry, particularly in the United States, which had only just begun its late entry into radio astronomy. At the time, the US Smithsonian Astrophysical Observatory (SAO) was able to conduct only optical tracking. The *New York Times* noted:

> No US Confirmation: No United States sources were able to confirm the reported speed-up in the circuit of the satellite rocket around the earth. The Smithsonian Astrophysical Observatory in Cambridge, Mass., which is in charge of visual satellite tracking in this country, confirmed receiving the report from Britain.[47]

Importantly, it was not only the Americans who were unable to keep track of Sputnik's rocket. Leaders in Moscow reached out to Lovell a couple of weeks after the launch, asking him to repeat the tracking to locate the rocket, because they too had lost track of it: at the time, the Soviets also did not have the appropriate radio telescope-tracking facilities. Because of his verification and promotion of Soviet success to the western world and aid with spacecraft tracking, Lovell became a figure of great renown in the Soviet Union. A few months after Sputnik's launch, he received a friendly New Year's telegram from the Soviets,[c] and they sent him subsequent telegrams that listed radio frequencies and predicted impact times for their future space missions, including Luna 1 in 1959. The Soviet Space Program began to view Lovell as an "unofficial 'record keeper'" that could verify its accomplishments for the United States and Europe, which, as noted earlier, were otherwise skeptical of Soviet achievements.[48]

Lovell's success at radio space tracking initiated Jodrell Bank's role in conducting secret military operations. This role included serving as part of the British ballistic missile defense system,[49] which often meant collaborating with defense operations in the United States. Like the Soviets, the United States approached Lovell in 1958, claiming they had "successfully tested their first intercontinental missile . . . but had no means of tracking it."[50] They requested that Jodrell Bank assist in this tracking, and Lovell complied. This set a precedent for future military involvement between the United States and Jodrell Bank. Jodrell Bank was thus put in a strange position. For several years during the early Space Race and Cold War period, it became the de facto space-tracking facility for both Soviet and American missile launches and space missions, since neither had yet developed the ability to track its respective payloads.[51]

In part as a result of his popular reputation in the Soviet Union, in 1963 Lovell was invited to visit Moscow to give lectures, and also to visit the Soviet's new radio facilities in Crimea. In a book titled *The Story of Jodrell Bank* that he published in 1968, five years after the trip, Lovell blithely described in some detail the visit and the background leading up to it, including the use of Jodrell Bank in aiding Soviet space tracking. Decades later, however, a collection of Lovell's notes, diaries, and other writings were donated to the University of Manchester's John Rylands Archive and released upon Lovell's death in 2012. His tone regarding the trip differed drastically from the benign account in *The Story of Jodrell Bank*. In his memorandum for the 1963 file, Lovell began forebodingly: "in retrospect I should not have accepted that invitation to visit the USSR from the Academy of Sciences in the summer of 1963 and further-

---

[c] The celebration of the New Year was important in the Soviet Union, and the exchange of New Year's cards, like the exchange of Christmas cards in the United States or in England, was a symbol of good wishes and friendliness. During my archival research I spotted several New Year's greetings cards from Soviet scientists to their friends and colleagues in the West.

Figure 2.6 This is an image of the ADU-1000 South Station transmitting array, part of the Yevpatoria Pluton complex in the Crimea. The Pluton complex supported the Soviet space programs as well as military activities. Credit: Rumlin / Wikimedia Commons.

more the Joint Intelligence Bureau in London should have advised me not to visit Moscow."[52] The jarring disconnect between Lovell's largely positive description in his 1968 book and his postmortem memorandum highlights how radio astronomy facilities played a dual role for both science and diplomacy.

Lovell's trip to the Crimea included a visit to the recently completed Yevpatoria facilities. As we've already seen, there was a strong connection between the development of radio astronomy facilities and the military – from the radar origins of much radio astronomy equipment to the potential for military application of radio astronomy techniques and expertise in intelligence gathering and telemetry. There is perhaps no example more illustrative of this intersection than the Yevpatoria Pluton Deep Space Tracking Station's ADU-1000 South Station transmitting array, built in 1961. The telescope array (Figure 2.6) had dishes welded onto the hulls of decommissioned World War II submarines and bridge trusses and was steerable thanks to being mounted on turrets from former battleship guns.[53] It quite literally embodied the relationship between the military and astronomy. Lovell was likely the first westerner to ever visit the complex that, as we shall soon see, became a point of obsession for US and British intelligence operations.

* * *

According to his memorandum and diary, Lovell's trip was superficially cordial and pleasant, but there existed an undercurrent of tension and paranoia. At one point, Lovell recounts being approached during a party by a man he did not know. He writes:

> We moved in the darkness from the blazing light. "What do you want?," I asked this man whom I have never seen before. "I must ask you to take immediate action to arrange for Professor Shklovsky to leave the country because he is in great danger" – "I have no power to do this" – "Not so, your authority is greatly respected in this Country and I must ask you immediately to return to England to ask the Vice Chancellor of your University to write to the President of the University of Moscow to extend a pressing invitation to Shklovsky to visit your Observatory."[54]

In his memorandum, Lovell noted that he was already familiar with Shklovsky. In 1963 Iosif Samuelovich Shklovsky was a professor of astrophysics at Moscow State University's Sternberg Astronomical Institute and was attaining fame, both within and outside the Soviet Union, as a foundational figure in the burgeoning field of radio astronomy. Before that, he had trained as a physicist at Moscow State University in the late 1930s, then had been sent to Ashkhabad during the war.[55] After the war Shklovsky returned to Moscow, where he led the Soviet effort to develop postwar astronomy. He published his first radio astronomy paper in 1947.[56] According to his colleagues, in 1952 Shklovsky "gave the world's first series of [university] lectures entitled 'Radio Astronomy'" and quickly became the leader of a radio astronomy group at Sternberg, which was a division of the University.[57] He later became a key figure in the Soviet Space Program, and in 1967 founded the Astro Space Center at the Lebedev Physical Institute of the Soviet Academy of Sciences.

Shklovsky had a reputation for thinking up ideas and theories that were sometimes groundbreaking, other times absurd. Of this contradictory aspect of his genius, one colleague wrote: "Fifty percent of Shklovsky's ideas are brilliant, but no one can tell which fifty percent they are."[58] To give an example of one of his impressive ideas, after the publication of Henrik van de Hulst's theoretical paper that predicted the emission of neutral hydrogen, Shklovsky independently calculated the intensity of the 21-centimeter hydrogen line – a breakthrough achievement for observational radio astronomy (for more explanation on the significance of this achievement, see Chapter 4). To give an example of a perhaps lesser idea, Shklovsky repeatedly published theories that one of Mars's moons, Phobos, was an artificial and hollow satellite placed in the Martian orbit by extraterrestrials.[59] This latter example illuminates another facet of Shklovsky's personality: his interest in CETI.

Following the publication in 1959 of the Morrison and Cocconi paper, which proposed the use of the hydrogen line for communicating with extraterrestrial intelligence, Shklovsky began to harbor an intense interest in the subject and supported his students in pursuing CETI projects. According to his diary, Lovell's relationship with him at this point largely consisted of "conversations about the problems then facing astronomy about the nature of the distant radio emitting objects in the Universe" – objects that would later be identified as "quasars."[60] In the Soviet Union such objects were then considered candidates for CETI searches, though it is unlikely that Lovell knew this at the time.[61] Despite his familiarity with Shklovsky and his interests, Lovell nevertheless recorded in his memorandum that he did not know why Shklovsky's life would be threatened.

The intersection between radio astronomy and military infrastructure created difficult and paranoid interactions between western and Soviet scientists, hindering interterrestrial scientific communication. The shadowy interaction with the unknown figure who pleaded for help to save Shklovsky's life clearly impacted Lovell's state of mind, and as his trip to Moscow wore on he, too, grew paranoid. At the end of Lovell's visit, the president of the Soviet Academy of Sciences, Mstislav Keldysh, told him that he knew about his efforts to build a larger telescope at Jodrell Bank and inquired about the cost. When Lovell remarked that it would be a challenge to obtain the £4 million required, Keldysh allegedly interjected:

"but that is only a very small sum in my budget. We would be very glad if you would stay in Russia and we will build the telescope for you."[62] This offer was a clear indication of the level of respect afforded to Lovell by the Soviet Union – and perhaps also an attempt from the United States to appropriate talent, since the United States, too, considered Lovell and his telescopes valuable for the conduction of telemetry.

Lovell refused and returned to England. In his memorandum, he revealed that he became "mysteriously ill" shortly after his return and, though he did not expand on his symptoms, he noted that the "intelligence agents" thought that the Soviets had used "some means (probably radiation)," perhaps from the radar beam of the telescope, to brainwash him or erase his memories of Yevpatoria.[63] The reason for doing so? Lovell speculated that the Soviet Academy had expected that he would accept its offer to build him a telescope if he relocated to the Soviet Union and, upon his decline, attempted to prevent Lovell from returning to the West and sharing news about the Soviets' new facilities in Yevpatoria.[64]

Lovell's accusation of assassination or brainwash by telescope was a bold claim, and the Rylands Archive snidely noted in its special collections blog that "already the diary has attracted a lot of interest from the media, who [sic] never allow the absence of hard facts to get in the way of a good story."[65] Lovell's claims certainly sound like a tagline from a Cold War science fiction film: did the Soviets attempt to irradiate Sir Bernard Lovell's brain using their radar beam? To give his claim due diligence, however, it is certainly *possible* that the Soviets tried to harm him, and the World Health Organization has information on the detrimental effects of radar to the human body.[66] I believe, however, that an attempted brainwash by telescope was highly unlikely.

I would argue instead that this recounted episode is emblematic of a larger trend in Anglo-American relations with the Soviet Union: fear of the technologically advanced unknown. Since the Soviet Union was a closed society, there was much speculation on what technologies it may have developed and for what purposes. As many science and speculative fiction scholars have noted, the 1950s' and 1960s' literature and film often represented Cold War anxieties about technology and domination by foreign civilizations under the form of invasions from ray gun-armed alien monsters and mind-controling weapons.[67] In the Introduction, we learned that science fiction during the Cold War period was an ideal medium through which to identify particular scientific–technical fears and concerns; and this includes brainwashing. One classic example comes from the film *The Manchurian Candidate* (1962), which was released the year before Lovell's trip. The film starred Frank Sinatra and Laurence Harvey, who plays a US soldier returned from the Korean War brainwashed by communists into becoming a "sleeper agent" – someone who assassinates targets unwittingly, through communist control.[68] Perhaps the recent release of *The Manchurian Candidate* affected Lovell's perceptions of the events he experienced.

Yet it was not only science fiction that revealed these fears; paranoid speculation on Soviet technological capabilities infiltrated even the upper echelons of western governments. Government activities fueled science fiction and science fiction similarly influenced governments, in a vicious cycle. *Masters of Deceit*, by J. Edgar Hoover, the director of the FBI, warned of mind control by communists. In this book Hoover dramatically cautioned that "many well-meaning citizens, attracted by these words and not seeing behind the communist intentions, have been swept into the communist thought-control net."[69] This fear of mind control or brainwashing was quite common. In fact historians have unpacked the history of psychological warfare during this period, showing that the Cold War was fought just as much in cultural, informational, and psychological campaigns as it was on military and technological battlegrounds.[70] Of course, Hoover's thought control net and the scholarship on cultural warfare refer to psychological tactics to control the mind – not quite comparable to the high-tech attempt conjured by Lovell. Fear of possible mind-controlling or mind-erasing weapons, however, did prompt the CIA to test for itself other high-tech psychological interventions, which culminated in projects such as MKUltra – a program of experiments operated by the Office of Scientific Intelligence of the CIA together with the US Army Biological Warfare Laboratories from 1953 to 1973.[71] MKUltra studied, among other things, the effects of certain drugs (e.g. LSD) on memory, coercion, and hypnosis.

Lovell's claims are also somewhat reminiscent of what we now call "Havana syndrome": a disputed medical condition used to describe a series of health incidents that affected US and Canadian diplomats and intelligence officers stationed in Havana, Cuba, from late 2016 on but that later were also reported by diplomats and officers in countries around the world. Various theories have been proposed to explain the cause of Havana syndrome, and these include targeted sonic or microwave attacks, much as in Lovell's claims. After thorough government investigation, and especially after a report released by the US government in March 2023, "most [Intelligence Council] agencies have concluded that it is 'very unlikely' a foreign adversary is responsible for the reported" symptoms of Havana syndrome.[72] The report states that "symptoms reported by US personnel were probably the result of factors that did not involve a foreign adversary, such as preexisting conditions, conventional illnesses, and environmental factors."[73] It is possible that the high stress experienced by diplomats and intelligence officers caused psychosomatic reactions. As with those afflicted by Havana syndrome, it is unlikely that Lovell was truly the target of brainwashing by telescope radiation. Rather his fears, along with the concerns of British intelligence, were representative of a culture of paranoia regarding technology and the unknown alien civilization of the Soviet Union. That being said, new evidence for Russian "energy weapons" is currently emerging and the debate on the source of Havana syndrome wages on. Perhaps Lovell's story fits into a larger narrative on Soviet–Russian covert attacks, or simply serves as another piece of evidence that high-stress environments can have detrimental mental and physical effects. Regardless, it is clear that Cold War radio astronomy operated in a tense and paranoid atmosphere.

The Longest Search

In the 1970s and 1980s there was a shift in how CETI was treated by the wider scientific community in the United States. Before then, US CETI consisted largely of targeted searches conducted by individual astronomers (as opposed to concerted efforts made by research groups or institutions), and very few observational studies were carried out in the United States by comparison with the Soviet Union in the 1960s.[74] The shift from mostly theorizing to planning and organizing a large-scale formal search in the 1970s also marked the departure from CETI to SETI – the *search* for extraterrestrial intelligence. As part of this US shift of focus toward searches, in 1971 Barney Oliver, a SETI scientist and the vice president of Hewlett-Packard, conducted a NASA-funded investigatory study titled Project Cyclops in which he proposed the design of a phased array for SETI.[75] Jill Tarter, who, as we know, later on developed the concept of a "cosmic mirror," was then a graduate student. When she read the Project Cyclops report, she was inspired to pursue a career in SETI and became one of the next-generation of scientists in the radio search for extraterrestrial intelligence.

Tarter and her fellow SETI colleagues were successful at integrating SETI not only into radio observatories (where the field had largely remained in the 1960s) but into scientific institutions such as NASA. The involvement of NASA and its comparably large budget made large-scale searches possible, and in the late 1970s NASA had its first opportunity to plan and design a NASA SETI project. Working at NASA's Ames Research Center, Tarter and her colleagues hoped to use NASA's Deep Space Network, which maintained 34-meter telescopes primarily designed for tracking deep space probes such as those used for the Pioneer and Voyager missions. In order to use the telescopes for their intended purpose, the SETI team had to design and build a prototype for a "multi-channel spectrum analyzer," which would be used to conduct (1) a targeted search of stars selected as best candidates for SETI and (2) an all-sky survey that would sweep the sky, looking for artificial signals.[76] Scientists at another NASA site, the Jet Propulsion Laboratory (JPL), wanted to participate, and so the project created two arms of the program to develop analyzers. The Ames prototype was called "Peterson's Left Leg," a humorous reference to Al Peterson, an electrical engineer at Stanford who provided technical support for the design.[77] The JPL prototype was simply called the MCSA – the multi-channel signal analyzer.[78] The analyzers were designed to search for "obviously engineered" signals, which the SETI group decided should be compressed in frequency to bandwidths that were narrower than whatever is possible in the natural astrophysical environment.[d] Even when searching for emissions from naturally narrowband astrophysical sources such as masers,[e] the group

---

[d] This was a somewhat arbitrary decision. There have been arguments in support of broadband searches as well and, as Ken Kellermann once wryly informed me, "unfortunately there is no cosmic treaty that specifies bandwidths for communication."

[e] A cosmic maser is a naturally occurring, single-pass amplifier that works at radio wavelengths, on the basis of the mechanism of stimulated emission. In simpler terms, it is a "laser" that is visible at lower frequencies than those our eyes can detect. In funnier terms, it's like an invisible rock-and-roll show that occurs around black holes. Space is cool.

did not know of any naturally occurring cosmic structures that were any more narrow than 300Hz, so the "sandbox" for SETI was defined at signals under 300Hz. The team developed an instrument that could observe about 65,000 individual spectral channels simultaneously, scanning a wide range of frequencies and identifying narrowband signals. Tarter later noted that this was "the type of instrument astronomers don't build, because nature doesn't do narrowband. So this was a unique instrument."[79] In other words, the analyzers were perfect for finding artificial signals, not natural astrophysical ones.

By the early 1980s, the JPL team was further along in its design than the Ames team. The MCSA fit inside a van, making it relatively transportable. Because of this portability, Sam Gulkis, JPL's project scientist, informed Tarter that he and the JPL team had been invited to field-test their equipment at Jodrell Bank. This would be a huge boon for the prestige of SETI. Even in the 1980s, when it was overtaken in scientific achievements by the next generation of radio telescopes and arrays, the observatory at Jodrell Bank still carried the prestige of its early role in the Space Race, and Bernard Lovell still helmed the director's office. Tarter later recounted her excitement that SETI was being recognized as important science by one of the world's most prestigious radio astronomy observatories, and "invited" herself along on the trip.[80] When she arrived in Britain, she learned that JPL would be testing the all-sky function, and so she decided to spend her time not with them, but instead testing the MCSA's capabilities for targeted searches by observing hydroxyl (OH) masers. It was during this time that Tarter, in a private conversation with her JPL colleagues, learned the true nature of the field test: to use the MCSA to conduct an intelligence search for a secret Soviet signal. Tarter's frustration with the situation was still palpable nearly 40 years later:

> I was incensed! Here I was, so proud, that Jodrell Bank is allowing us to... ugh! I was so pleased and proud of myself, that here we are doing field tests for SETI at Jodrell Bank, isn't this fabulous, and then indeed it was just a big shill, it was just a cover story. So I was really furious with my colleagues. I just... it was... [Tarter groans]. I can't... I don't know how to express it. I was so... again, very young, very naïve, very full of myself thinking that SETI was so important... and that it would obviously be invited to do a field test

at Jodrell Bank, you know? Because we were so special and so good! And, as it turned out, it wasn't like that at all and it took me a while to calm down.[81]

A report from the NSA sheds light on what had occurred. The report, declassified in 2011 and titled "The Longest Search," was published in the *Cryptologic Almanac*, a classified academic journal published internally by the NSA. It began by noting the long history of searches for particular signals known to the intelligence community, but not yet found and identified. One of these "white whales" was the Soviet deep space data link, which was used by Soviet space probes such as the Venera missions to send images from Mars and Venus back to the Earth (among other things).

The historiography of signals intelligence has been described by historians Matthew Aid and Cees Weibes as an "inventory of ignorance," given the enormous lack of information and publication on how the United States conducted signals intelligence collection during the Cold War era.[82] As we saw with the Sugar Grove listening station, SIGINT and communications interception were significant parts of US Cold War efforts to gather information on the Soviet Union. Yet much of the purpose behind SIGINT is often concealed, sometimes even from intelligence officials themselves. Even the Longest Search report questioned the obsession with finding the Soviet deep space data link, noting that, though the search for the signal began in 1962, shortly after the first Venera mission launched in 1961, the signal evaded analysts until 1983. Given that most SIGINT searches only last for a few years, the report claimed: "if this search was not quite a Moby Dick-like obsession, it never entirely left the minds of those analysts who wanted the signal, either."[83] The Soviet deep space data link consisted nearly exclusively of radar mappings of planets, images from the surface of Venus, and telemetry data. This information was valuable to the scientific community, of course, but the images of Venus' surface and other scientific data were nearly always released by the Soviet Union shortly after they were received, as part of the country's aim to establish its position as the dominant space power. There does not appear, then, to be a clear motivation, either scientific or political, for the effort put into capturing the signal. The NSA report concluded by acknowledging the futility of the search: "in the final analysis, though, there seems to have been few obvious benefits from this prolonged search for the Soviet deep space data link."[84] Nonetheless, it is clear that, if there was not a strategic benefit to the search, there was certainly an emotional one, perhaps especially on the part of Lovell, who, as we have seen, feared the Soviet Union and prided himself on the historic role Jodrell Bank played in the Cold War.

But why did Lovell need to use the SETI van, as opposed to existing SIGINT tools and techniques, to capture the Soviet deep space data link? C/SETI scientists and engineers had great interest in identifying artificial sources, and this was recognized by the intelligence community. In fact there had even been some courting on the part of the SETI community; Jill Tarter explained that two SETI scientists, Kent Cullers and Carl Sagan, visited the NSA on an "information exchange," to give a presentation on the tools developed by SETI, but also in the hope that they might learn from the intelligence community as well. As Tarter rightly noted, "we had gotten a lot of technologies that we use in astronomy today out of [the] military, classified development that then gets unclassified."[85]

This phenomenon is also true in the reverse: intelligence agencies sometimes benefited from the development of scientific instruments. In their pursuit of the Soviet deep space data link, CIA hardware specialists would attend international space exhibitions to investigate waveguides on Soviet equipment and "discovered that the equipment was configured to transmit a signal somewhere between 5.6 and 6.3 GHz," homing in on the frequencies the link might have used. Furthermore, astronomers themselves got involved – an example of human intelligence (HUMINT). According to the NSA report, "Western astronomers who were aware of the search for the missing data signal discreetly queried their Soviet colleagues about the Soviet data link. One was told that it was 5.9 GHz."[86] Still, after decades of searching, these forms of intelligence gathering were clearly insufficient for capturing the link.

The search was further complicated by geopolitical conflict. The NSA report notes that, because of geopolitical instability (most notably civil war in Ethiopia), the United States had lost control of intercept sites in Turkey and Ethiopia, which meant that it "could intercept transmissions only during [a] short window."[87] The portability of the RFI van made it a valuable asset, and the NSA report described this van as "a system designed specifically for the collection of signals from deep space . . . it was a unique configuration of receivers, spectrum analyzers, and computers" that "included a digital signal analysis subsystem that could monitor 64,000 radio channels, each 205 Hz wide simultaneously."[88] The report does not refer to Lovell by name, and the references to Jodrell Bank are redacted in the declassified document. It does, however, highlight that the use of the SETI technology was a resounding success. Because "the SETI specialists were given sanitized search parameters and limited feedback on results," the "true" purpose of the visit was conducted surreptitiously, and "shortly after midnight on 9 November . . . the 21 year search was over."[89] The document optimistically notes the usefulness of SETI technology, arguing that it "pointed the way to the advanced collection and signal analysis systems," perhaps with "some application to the study of Soviet space communications, especially with its constellation of intelligence satellites that circled the earth."[90]

While the intelligence community was satisfied, Tarter was furious. She recalled that, when she learned that the MCSA was being used by Jodrell Bank for intelligence gathering purposes, she attempted to angrily confront Lovell in his office. When she arrived, however, she found him standing over a book on his desk and weeping. Tarter described herself as "just overwhelmed" by finding the eminent scientist in such a state and discovered that he had been examining pictures of Dresden in World War II. He told her: "My sister tells me that I should be ashamed of myself."[91] The radar he developed during the war had allowed for successful bombing raids even through bad weather.

This story underscores the complexities of twentieth-century radio astronomy, as it evolved from the tragedies of World War II and the tensions of the Cold War. Some practitioners, such as Tarter, were unhappy with the field's entanglement with military applications. She would later develop the concept of the cosmic mirror, which posited that SETI was a force for peace and global unity. Yet the irony remains that the very tools designed to search for extraterrestrial intelligence were repurposed for espionage and conflict – searches for terrestrial intelligences. This juxtaposition highlights the enduring struggle between the pursuit of knowledge and the machinations of war, a dichotomy that continues to shape the legacy of radio astronomy and C/SETI.

3

Telegram Killed the Radio Star
The First "False Alarm" in CETI

CETI was a science built on analogy. Analogues tend to be an important feature of most of the space sciences. Astronomers cannot perform experiments on heavenly bodies; in consequence they often rely on analogues such as computer simulations. Astrobiologists study Earth analogues, for example extremophiles (i.e. microorganisms that live in extreme Earth environments), which help them consider the variety of life that might exist on other worlds. Aerospace engineers cannot use the Martian landscape to test their rovers, hence they use as analogues extreme Earth environments such as the Atacama Desert in Chile.

But CETI was dependent on analogues and analogy more than any other space science; after all, it is in some ways a science without a subject. At the time of the publication of this book, there is no empirical evidence of the existence of extraterrestrial intelligence. So CETI scientists creatively employed analogies in their search, to devise strategies and prepare for the possibility of contact.

One popular analogy in CETI history has been the one between contact scenarios on Earth and their possible counterparts in outer space. SETI anthropologist Kathryn Denning has shrewdly noted that "every time we tell a story about what will happen in the event of a detection, or contact, we retell the story of contact here on Earth."[1] This is another way of referring to the CETI concept of a cosmic mirror, which posits that, when we consider extraterrestrial intelligence, we are inevitably reflecting back on our own civilization. A common analogy in the early period of CETI was between extraterrestrial encounters and the kind of contact that Columbus had with Indigenous peoples during his colonial voyage, in what we now call the Americas. CETI scientists looked to Earth's past to speculate on what might happen when one civilization (especially a technologically 'advanced' supercivilization) encounters another; and they drew heavily on the Columbian Exchange when doing so.²

But as Denning has noted, "analogy carries its own baggage."³ She argues that analogy in C/SETI has great limitations, stating that "when we use one thing as a basis for understanding another, this can illuminate but also cast shadows and confound."⁴ In Chapter One, I mentioned the historian William McNeill and noted how CETI scientists expected him to make predictions on extraterrestrial civilizations rooted in his understanding of contact between Earth civilizations in human history. McNeill was frustrated by the analogies, arguing that the nuanced narratives found in the discipline of history cannot be deterministically projected onto worlds and beings we do not even yet know exist. Analogies in CETI are clearly limited in their use.

Nevertheless, I open this chapter by reminding you of a key analogy we'll use throughout this book: in many ways, the United States' and Soviet Union's attempts at communicating with extraterrestrial intelligence mirrored their attempts to communicate with each other. There are two good reasons why this analogy should not be dismissed the way we might dismiss others. First, because it is an analogy that CETI scientists themselves used to describe their efforts. As we shall see, CETI scientists were highly aware of the parallels between the challenges posed by communicating with extraterrestrials and those they were experiencing in trying to work with one another across the Iron Curtain. And, second, because although the Cold War made it difficult for astronomers in the United States and Soviet Union to work with one another, it also created a motivation to do so anyway.

The tensions between radio astronomy's role in both international collaboration and international conflict become most evident when we explore the history of attributions of discovery in radio astronomy. Internationalism and the universality of scientific priorities constitute a prominent ideology in radio astronomy, and this mentality certainly promoted and facilitated international collaboration even during geopolitically contentious periods. That said, as one delves into the history, it becomes clear that this closely held ideology did not work in practice as well as it did in theory, because national and military motivations interfered with scientific goals and achievements. The entanglement of radio astronomy with the military often hindered communications between scientists across countries – scientists who, ideally, would cooperate to achieve their professional goals. This situation often led to misattribution of discoveries and to international embarrassment for both states and scientists.

Attribution and Discovery

I once attended a meeting of the Historical Astronomy Division (HAD) of the American Astronomical Society, the largest society of professional astronomers in the United States. During the meeting Kevin Krisciunas, the HAD chair at the time, presented a slide that stated: "One could consider that the purpose of the Historical Astronomy Division is to give credit where credit is due for discoveries of the past." As a historian of astronomy, I believe that the history of this discipline serves a greater purpose than assigning credit alone, but Kevin was certainly correct in noting that a system of establishing credit for discoveries is a critical part of most sciences, and astronomy is no exception. But barriers to communication on Earth have sometimes meant that assigning credit was more easily said than done. This was especially the case during the Cold War.

There are several instances in the history of astronomy where the Soviets made a discovery before the rest of the world but, as a result of the communication barriers caused by the Cold War and the isolation of the Soviet state, it was western scientists who often received credit for the discovery in question, even if they made it at a later date. For example, Soviet astronomers at Pulkovo Observatory and at the Lebedev Physical Institute discovered radio recombination lines as early as 1963 or 1964.[5] But, given the lack of clear and consistent communication across borders, the credit for the discovery of these lines is usually assigned to Bertil Hoglund and Peter Mezger, who reported their own independent discovery in 1965, using observations made at the National Radio Astronomy Observatory (NRAO).[6] What is more, while radio astronomy observatories in the United States and in most other western countries were civilian organizations in spite of having some connection with the military, Soviet radio astronomy facilities were often run by the military and therefore many of their publications were heavily redacted or censored, which made it difficult for western scientists to verify their scientific validity.[7]

This communication barrier presented a challenge to astronomers because of an inconvenient truth that I touched upon when I described Lovell's desire to have the Americans pick up the tracking of Sputnik's rocket: the Earth rotates. If they wish to have continual observation or verification of their observations, astronomers must rely on their colleagues to observe the sources when these leave their own field of vision. It is for this precise reason that a system for astronomical telegrams was established. When something new was discovered in the sky (say, a comet or a stellar outburst), this did not mean that it would necessarily stick around for long; hence it necessitated quick verification by other observatories or prolonged observation.

To solve the problem of fast, worldwide communication between astronomers, the Central Bureau for Astronomical Telegrams (CBAT) was founded in 1882. This was a direct result of the crisis of communication during the Great Comet of 1882, when astronomers who tracked the comet realized that their informal astronomical news transmission systems were "thoroughly inadequate."[8] In the early twentieth century, the International Astronomical Union (IAU) designated the CBAT as its official telegram bureau and created a commission to oversee its management.[9] A 1968 article in *Physics Today* boasted that over 600 astronomers subscribed to CBAT telegrams, as well as over 100 observatories.[10] In addition to facilitating communication and collaboration between astronomers at great distances, CBAT filled another purpose: it allowed astronomers to claim discoveries as their own to the international astronomical community. Comets, for example, are often named after the person who first discovers them. Thus, if an astronomer was to observe a new comet and quickly send out a telegram to CBAT, they would have a good chance of earning credit as the first discoverer, in which case the comet would be named in their honor.[11] Additionally, astronomers would report their findings so that they could be verified by other astronomers, and this would secure their discovery as a legitimate one. Telegrams often included the coordinates, the date and time of the initial observation, and other such information. Sometimes they were very short, with just a few lines of data, other times they contained information and citations to contextualize the discovery. CBAT set precedence for this type of communication system, and over the course of the twentieth century more circulars and bulletins were created. By the start of the Cold War period, the IAU astronomical telegram had become a common form of communication between astronomers around the world.

Despite these efforts to promote ease of communication between astronomers around the world, radio astronomy's ties to the military, especially in the Soviet Union, presented challenges to the efficacy and transparency of the communication system. Consider the case of the first reported observation of intrinsic, periodic, extragalactic radio variability. Variable stars are stars that have brightness that fluctuates when viewed from Earth, meaning that they may appear to "blink," fade, or grow brighter over time. This can happen if a star is surrounded by a disc of dust that blocks some of the light coming from the star, altering its apparent brightness. *Intrinsically* variable sources change not because they are obscured by dust or by an orbiting object, but because of changes in the physical properties of the star. This happens in certain circumstances, for instance when stars give off flares and mass ejections, or when a star expands or contracts in its evolutionary lifecycle. By the mid-twentieth century there were many observations of optically variable sources; in 1961 the IAU even launched an entire astronomical bulletin dedicated to telegrams – the *Information Bulletin on Variable Stars* – to which astronomers could send their data on new variable stars. Within radio astronomy, however, it was thought that radio sources that were strong enough to be observed by the low-sensitivity instruments of the mid-twentieth century only existed at great scales and distances. As noted earlier, many optical astronomers considered the radio universe to be far less interesting than the optical; it was not even certain whether intrinsic, periodical, variable radio sources existed at all and, if they did, whether they could be observed at human lifetime scales. By the early 1960s, the only intrinsically variable extragalactic radio source that had been observed was Cassiopeia A, the remnant of a massive supernova, which slowly faded over time – a far less exciting object than the periodically variable sources found at optical wavelengths.[12] The radio universe was thought to be a quieter, less dynamic place.

In August 1964, however, a Soviet graduate student named Gennadii Sholomitskii began to use the deep space tracking antenna in Yevpatoria to observe a radio star called CTA-102.[13] The term "radio star" referred to sources that were believed to be stellar objects that produced copious emissions of various radio frequencies. Sholomitskii claimed to have

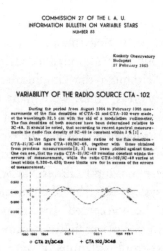

Figure 3.1 The IAU telegram of Sholomitskii's results, showing the variability he observed in CTA-102. Credit: Sholomitskii, G. B. "Variability of the Radio Source CTA-102." *Information Bulletin on Variable Stars* 83 (1965). IAU Commission 27.

discovered radio variability in the source "with a period of about 100 days" in CTA-102 (that is, the object appeared to "blink" across the span of a few months).[14] To put his achievement in context, astronomers would have been stunned to come across an object that varied over a period of years, let alone months. Sholomitskii's discovery had the potential to make a great impact on the field, transforming astronomers' understanding of the radio universe (and saving all those jobs at NRAO). In 1965 he announced his discovery through a telegram submitted to the *Information Bulletin on Variable Stars,* which quickly gained attention from astronomers in the West (see Figure 3.1).[15] He also published a more detailed article on his discovery in the Soviet *Journal of Astronomy – Astronomicheskii Zhurnal* [Астрономический журнал] – which during the 1960s was translated and republished in the United States as *Soviet Astronomy*. Soviets excited by the discovery called the strange blinking source the "lighthouse in the sky."[16]

Given the secret nature of the facilities at Yevpatoria, many of the details of Sholomitskii's observation were left unpublished for security reasons, which made it difficult for scientists in the West to verify his results.[17] Sholomitskii's telegram was upheld as an example of why Soviet science was untrustworthy and further cemented a culture of distrust between scientists on opposite sides of the Iron Curtain. In fact, the American astronomer Kenneth Kellermann mentioned earlier was one of those who initially doubted Sholomitskii's claims. In a 1968 paper Kellermann stated that, although the "implications of Sholomitskii's discovery were clearly very great," there were many theoretical problems with that discovery.[18] Besides, no observations of this variability had been made by western astronomers, despite the fact that CTA-102 was a relatively well known object.[19] Kellermann argued that, because of the secretive nature of Soviet observing facilities, "little is known about the antenna or radiometer system used by Sholomitskii," and hence "his results have generally not been accepted" by astronomers in the United States.[20] It was not until radio variability became generally accepted several years later, thanks to observations of other sources reported by western scientists, that it was confirmed that CTA-102 does indeed vary at the scales reported by Sholomitskii.[21] But by that point the study of radio variability was firmly established in the field, and Sholomitskii had left radio astronomy for infrared astronomy.[22] Sholomitskii is hence not a very well-known name in the history of radio astronomy.

Of course, Sholomitskii's observations constitute but one out of dozens of similar cases that occurred during the Cold War. What makes this case significant, however, is that it demonstrates not only the connection between radio astronomy and the military but also the connection between CETI and interterrestrial crises of communication. In fact I have already discussed CTA-102 in this book – not directly, but in discussing Bernard Lovell's relationships with Soviet scientists. In Chapter 2 I referred to conversations between Lovell and I. S. Shklovsky in the early 1960s in which they broached "problems then facing astronomy about the nature of the distant radio emitting objects in the Universe."[23] Shklovsky was Sholomitskii's professor; it was he who had instructed Sholomitskii to observe CTA-102 because it was one of those newly discovered objects and initially no one was quite sure what they were.[24] The label these objects were tentatively given was "quasi-stellar" or "quasar," because they displayed strong luminosity, like a nearby star, but had other confusing properties that made it difficult to ascertain whether or not they were small and inside our galaxy or large and terribly far away.[25] The first identification of these objects came in 1960, from the Owens Valley Radio Observatory at the California Institute of Technology

(CalTech). The initial aim of the observation that led to the discovery of these objects had been to observe faint radio sources, likely in the form of distant galaxies. It seemed unlikely, however, that the resulting sources were extragalactic (i.e. outside our galaxy), given their unrealistically high luminosity.[26] But in 1963, just a year before Sholomitskii's observations of CTA-102 began, CalTech's Maarten Schmidt realized that one of these so-called quasi-stellar objects, 3C 273, had a large redshift[a] of 0.16.[27] With further confirmation, this would come to mean that the sources in question were not "radio stars" in our galaxy, but some of the brightest and most distant objects in our universe – a discovery with important implications for astronomy. Yet, as we shall see, Schmidt's observations were not easily communicated to the Soviet Union, and this led to speculation in the Soviet Union that the source was actually a sign of extraterrestrial intelligence. The discovery of variability in CTA-102 was one of the earliest events in Soviet CETI history, but also led to one of the first "false alarms" in the history of the search for extraterrestrial intelligence.

The Birth of Soviet CETI

The United States is constantly messaged by aliens. Or rather you might believe that, if you were to take science fiction literally. Culturally, there are strong associations between extraterrestrial civilizations and the United States. This is partly due to the country's role in the Space Race, of course, but also to the fact that C/SETI's development in the United States has won it a place in the public consciousness. Carl Sagan is a household name in the United States. The Very Large Array (VLA) was made popular by a SETI-themed science fiction film starring Jodie Foster (the VLA now has a thriving visitor's center and is a popular New Mexico tourism destination). Many of the current large SETI projects are hosted by institutions in the United States. But, in its infancy, CETI was not a science dominated by the United States alone; it developed in

---

[a] In astronomy, "redshift" refers to the shift of light toward longer wavelengths (the red end of the electromagnetic spectrum) that is due to the expansion of the universe. A redshift value of 0.16 indicates that the observed wavelength of an object's light has been stretched by a factor of 1.16. A redshift of 0.16 corresponds to a distance of approximately 2 billion light-years. In 1960, this was one of the first objects to be identified at such a large distance.

the Soviet Union as well, around the same time and with similar fervor. But this is much less commonly known, even within the contemporary SETI community.

This gap in our knowledge is understandable. Partly it occurred because Soviet CETI largely fell apart around the same time the Soviet Union did, in 1991. But it also occurred because of the repression of information in modern-day Russia. The relationship between astronomy and the military, especially in the former Soviet Union, has shaped access to CETI's historical record, making it extremely difficult to give credit to Soviet CETI scientists, even in a historical capacity.

Visiting Russia for CETI research was quite challenging for me. I was able to do so only with the support of a global network of radio astronomers who assisted me (thanks to them, I was able to obtain a scientific visa). I had to receive a formal invitation directly from the head of the Russian Astro Space Center, Nikolai Kardashev, whom you will learn more about in this chapter. During my visit, I had the support and assistance of scientists housed in the Russian Space Research Institute and was able to conduct interviews with the many helpful and friendly astronomers there. But shortly after I left Russia, in October 2019, President Putin signed a foreign media amendment that used vague wording to demand that any individual, Russian or foreigner, who published information, in the media or online (on a blog, for instance), and who also received money from a foreign funding source be declared a "foreign agent," a loaded term with Cold War-era intelligence implications.[28]

The amendment was likely targeting journalists as part of the Putin administration's crackdown on the free press, but it inadvertently affected historians (such as myself) who publish work through overseas scientific and academic institutions and receive funding or grants from non-Russian sources. If I am able to return to Russia someday, obtaining access to secure facilities such as the Russian Space Research Institute will be nearly impossible if I am labeled a foreign agent (and I have to assume that writing this book won't help). Furthermore, many Russian archives are closed to outside visitors. Despite my best efforts and attempts to pull strings through the many friends and connections I had within the Russian scientific community (including administrators of institutes, directors of observatories, and even cosmonauts), I was unable to access a single scientific archive during my time in Russia. I was forced to rely instead on oral history testimony to complement published sources and to establish some elements from the Soviet side of the historical record on radio astronomy; and I had to use Soviet sources that were preserved by US scientists, such as letters of correspondence with Soviet scientists held in the NRAO Archive.

Despite the internationalist rhetoric of CETI, its institutions were so entrenched in Cold War politics and military operations that this involvement has even interfered with historical research that took place decades after the Soviet Union dissolved. Furthermore, there have been few comprehensive English translations of key documents or papers from Soviet CETI and radio astronomy history. Many of these papers are also inaccessible from British or American libraries. I have been very fortunate to have colleagues with ties to Russia who have scanned and sent me such records, which are otherwise extremely difficult to obtain in the West.

It is important, then, to speak briefly to the limitations of this study. Given archival restrictions in former Soviet states and a relatively rich tradition of archival preservation in the United States, the historical record is extremely uneven between the two nations. As stated here, I have tried to rectify this imbalance by conducting oral history interviews with former Soviet astronomers. These interviews were supported by funds from the American Institute of Physics, then donated to its oral history archive. Nevertheless, oral history is an imperfect source; many of the men and women I interviewed were octogenarians trying to recollect events from 40 or 50 years ago. Where I could, I tried to match their claims with documentary evidence, and where I could not, I am sure to state that this is one individual's interpretation or recollection of events. There are, of course, some benefits to using oral history over archival sources. Soviet historian of technology Michael Gordin has noted that the history of technology in the Soviet Union has too often focused attention on the machinery and technology, without giving due attention to the human beings who create, operate, and interpret said technology. In writing about language and its intersections with technology in his analysis of Soviet machine translation, Gordin emphasized the need for the historiography of Soviet technology to "alienate the machine ... and return the human ... to our narratives of Soviet technology."[29] My approach to the history of Soviet technology takes Gordin's advice to heart; this book shows how technology shaped the science, but also highlights how both the militaristic and the internationalist goals of humans shaped the technology.

So what makes Soviet CETI distinct from US CETI? Drake's Project Ozma jump-started the field of CETI in the United States, but had influence in the Soviet Union, too. US CETI was partly inspired by the proliferation of space-based military signals, and this is certainly true of the Soviet Union as well. But US CETI is unique in the inspiration it drew from the Columbian exchange and the myths of first contact in the new world – an aspect I criticized in Chapter 1. In the Soviet Union, however, a different myth, one rooted in a pre-revolutionary philosophy, Russian cosmism, was likely a major contributor to the development of Soviet CETI.[30] Originating in imperial Russia, cosmism is a widely known philosophical and cultural movement in Russian academia, yet has gained little attention from western academics, given the dearth of translated sources.

Cosmism developed in the late nineteenth and early twentieth centuries, spearheaded by the philosopher Nikolai Fedorovich Fyodorov and inspired by science fiction and the expansion of rocketry for space travel.[31] The tenets of cosmism focused on "active self-directed human evolution; the need for universal solutions to existential problems ... universal immortality as a human task; and a view of man as a citizen not only of the earth but of the entire cosmos."[32] Proponents of cosmism believed that "the world is in a phase of transition from the 'biosphere' (the sphere of living matter) to the 'noosphere' (the sphere of reason)."[33] The results of this transition are what cosmists called "planetarian consciousness," which posited that humanity would overcome sectarianism and divisiveness and all humans would rationally work together to conquer disease and death, ending up as "an immortal human race," which would spread throughout the cosmos.[34] Crucially, cosmism was also technologically determinate. According to cosmists, the predestined "teleological evolution" that leads to cosmic immortality would be guided by technological development – it would be a technogenesis.[35] Since cosmism predicted a future in which all humanity takes to the stars to live out an immortal destiny in the cosmos, it was particularly appealing to those who had dedicated their lives to the sciences of space.

Cosmism's explicit role in twentieth-century Russia, however, was short-lived. After the communist revolution and the establishment of the Soviet Union, it was effectively banned on the grounds that it represented one of many forms of counterrevolutionary mysticism, which the Communist Party worked to suppress. The party fought an uphill battle against magic, religion, and occultism under the banner of its Marxist–Leninist atheist vision; cosmist ideologies were repressed for their spiritually charged undertone. Despite its official commitment to atheism, however, the Soviet Union played with various pseudo-religious forms, and its state-mandated mythology drew inspiration from the very cosmist thought it wanted to repress. Just as Americans believed in a cosmic and manifest destiny, Soviets too saw something of themselves in the cosmos. While Americans created narratives about a final frontier, an extension of the driving force of manifest destiny, Soviets admired what they saw as the inherent communism and alleged anti-imperialism of the cosmos, which was the province of humankind.

Many cosmist thinkers such as Konstantin Tsiolkovsky, the father of Soviet rocketry, fantasized about a human destiny spent among the stars and a utopian future of harmony and cooperation, not all that different from the CETI cosmic mirror. A sense of awe and wonder is fundamental to religious belief, even when this belief does not involve a personal god or deity; cosmism created similar theological sentiments in those who followed it. The Soviet Union may have banned cosmism, but the Soviet preoccupation with its main tenets, and especially with utopian futurism and cosmic ambition, remained not only in the popular zeitgeist but in the minds and practices of Soviet scientists. In fact several proponents of cosmism, especially the Russian philosopher and cosmist Alexander Konstantinovich Gorsky, believed that cosmism complemented Soviet goals and campaigned on its behalf, even writing to Stalin letters in support of it (a risky move, as we shall see in the next chapter).[36]

In Russian academia, CETI scientists are considered among those who promoted cosmist ideals during the Soviet era. In a 2012 article in the *Herald of the Russian Academy of Sciences*, Soviet CETI publications are upheld as complementing "the worldview of Russian cosmism" and presenting ideas that served as this philosophy's "theoretical cornerstone."[37] Further research is needed to fully understand the role of cosmism in twentieth-century scientific internationalism, but it is clear that futurist and expansionist philosophies were an important facet of the beliefs shared among US and Soviet space scientists, including CETI.

Cosmism was not the only philosophy that contributed to the development of Soviet CETI. Soviet science emphasized the importance of adhering to the principles of dialectical materialism to gain state support. In an article published in 1968, the Soviet astronomer Nicholas Bobrovnikoff claimed:

> [Soviet scientists] are emphatic that their materialistic philosophy is in complete agreement with the idea of extraterrestrial civilizations. According to this philosophy life is a normal and inevitable consequence of the development of matter, and intelligence is a normal consequence of the existence of life. Even the best-informed scientists in the USSR, like Oparin and Shklovsky, must necessarily subscribe to this [Marxist interpretation].[38]

The word "inevitable" is important here; like cosmism, communist ideology was in many ways determinate, purporting to offer an inevitable progression of society (toward communism). Even before the space age, this determinism was often cosmically oriented. Frederick Engels, Marx's close friend and collaborator (the two co-authored the famous *Communist Manifesto* in 1848), was a firm believer in the deterministic nature of historical development and the inevitability of proletarian revolution. Communism was a distinctly materialist philosophy, meaning that it aimed to explain the world solely in terms of tangible and physical factors, rejecting the existence of supernatural or spiritual entities. In other words communism was conceptualized as being scientific. First published in the Soviet Union in 1925, Engels' *Dialectics of Nature* applied to science the principles of dialectical materialism derived from Marxist theory. In that book Engels argues that the entire universe follows a biological destiny:

> However many millions of suns and earths may arise and pass away, however long it make take before the conditions of organized life arise, however innumerable the organic beings that have to arise and to pass away before animals with a brain capable of thought redeveloped from their midst, and for a short span of time find conditions suitable for life only to be exterminated later without mercy, we have the certainty that matter remains eternally the same in all its transformations, that none of its attributes can ever be lost, and therefore also that with the same iron necessity that it will exterminate on the

Earth its higher creation[,] the thinking mind, it must somewhere else and at another time again produce it.[39]

In summary, although intelligent life may arise and perish on Earth, the enduring nature of matter ensures that similar life forms will emerge elsewhere in the universe at different times. This ontology emphasizes the cyclical nature of existence and the eternal transformation of matter. In other words, Engels promotes the idea that intelligent life in the universe is normal, common, and – critically – inevitable. Clearly, both communism and cosmism laid the ground for the development of CETI in the Soviet Union, shaping its character and influencing ideas in the field.

The most comprehensive study of Soviet CETI/SETI to date is a paper published by two former Soviet astronomers, Leonid Gurvits and Lev Gindilis."[40] The article, written in the style of an astronomical review paper, charts the major CETI, SETI, and technosignature projects from the turn of the century to the present day. Gurvits and Gindilis credit a journal article from a former student of Shklovsky's, Nikolai Kardashev, as having "set the stage for [CETI] studies in the USSR."[41] This paper was published in 1965 in the Russian *Journal of Astronomy*.[42] Kardashev was then a young scientist working at the Sternberg Astronomical Institute in Moscow, in a research group under the leadership of Shklovsky. In his seminal paper, Kardashev was preoccupied with how to identify an artificial extraterrestrial source, should one be detected. He began by attempting to identify which wavelength range would be most suitable for interstellar communication. He did so by calculating the noise spectrum of our Milky Way galaxy, and found a deep minimum at the decimeter and centimeter wavelengths. This was significant because a search conducted in the minimum meant that the observer would get a greater signal-to-noise ratio for any faint signals (put simply, signals in this range would be easier to detect). A similar approach was taken by Drake in Project Ozma, and CETI scientists would continue to debate which wavelengths were ideal for searching for extraterrestrial signals (Figure 3.2).

The next issue Kardashev tackled was that of transmission power. In any given civilization, Kardashev posited, the rate of transmission of signals would be dependent on the amount of power available to that civilization.[43] To address this matter, he introduced his now famous

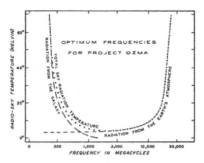

Figure 3.2 A graph from Drake's Project Ozma talk at the 25th Anniversary Green Bank conference, p. 20. Credit: NRAO Archive.

scale of technological civilizations, which is sometimes referred to as "the Kardashev scale."[44] The Kardashev scale outlined three types of civilizations that might exist in the galaxy; they were determined by their levels of energy consumption, a choice that had great implications for CETI conceptions of civilization. In his scale, type 1 was described as a civilization with a "technological level close to the level presently attained on the Earth."[45] Type 2 was a "civilization capable of harnessing the energy radiated by its own star."[46] In this part of the description, Kardashev made a reference to a "Dyson sphere," which was a theoretical technological concept envisioned by the British physicist Freeman Dyson. In postulating the "largest feasible technology" CETI scientists could look for, Dyson had imagined a technosphere built around a star and designed to exploit the star's entire energy output.[47] The concept became popular with CETI scientists interested in searches for passive signals, as the Soviets primarily were. Passive signal searches looked for extraterrestrial signals that were unintentionally observable from Earth; they did not contain a targeted message that tried to make deliberate contact. A passive signal could be radar activity from an alien airport, whereas a direct signal could be a beamed signal encoded with information and communications. Some Soviet CETI astronomers believed that looking for intentional signals would be a massive waste of time, like looking for a needle in a haystack (after all, you'd have to look in the right place at the right time and at the right frequency). In creating this scale, Kardashev attempted to argue that CETI searches should be dedicated to looking for "supercivilizations," whose radio activity would be easily noticeable, even if there was no direct message being sent toward Earth. He criticized attempts made in the United States, such as "the OZMA[b] [sic] project," which searched for type 1 civilizations, arguing that the possibility of detecting such civilizations was much lower than the possibility of detecting type 2 and type 3 civilizations.[48] Kardashev took Dyson's concept of large-scale technology a step further with his type 3 description: "a civilization in possession of energy on the scale of its own galaxy." Kardashev believed that such a civilization would be much easier to detect than an Earth-like one. His scale would shape the character of Soviet CETI.

* * *

After establishing his scale, Kardashev noted that the power outputs discussed in his paper (especially those of type 2 and type 3 civilizations) were "very close to the power of synchrotron radiation from nebulas formed in supernova explosions, or from radio galaxies."[49] Therefore, he argued, it was of utmost importance to distinguish between artificial and natural radio sources. Two of the criteria he outlined for potential artificial sources were a "very small angular [dimension] . . . less than 0.001 [milliarcsecond]" and the display of variability.[50] As already noted, in 1963 no intrinsically variable radio sources outside our galaxy had yet been identified, with the exception of Cassiopeia A, which was variable only in that its flux density decreased over time, as the supernova aged and cooled. An artificial source, on the other hand, might be irregularly or periodically variable as a result of intelligent activity or transmission of information by way of a modulated signal. Periodic variability, Kardashev therefore noted, was "obviously a criterion of outstanding importance" for determining if a signal was artificial.[51] Before concluding, Kardashev briefly mentioned that two sources that had been recently discovered in the United States at CalTech fit some of the criteria he had outlined. These sources, CTA-21 and CTA-102, displayed small angular dimensions[c] and exhibited "a spectrum highly similar to the anticipated

---

[b] Many in the Soviet Union did not understand Drake's reference to the character in L. Frank Baum's novels and thought that Ozma was an acronym (OZMA), as is usually the case with the names of many other US astronomical projects.

[c] It is important to note that, in the original Russian article, Kardashev writes СТА 21 и СТА 102 имеют угловые размеры менее 20 угловые секунды, which means "CTA 21 and CTA 102 have angular sizes of less than 20 seconds of arc." In a serious translation error, the English

artificial spectrum."⁵² At that time, cosmic radio sources were known to have a spectrum with lower flux density at high frequencies (which led to a steep spectrum); an artificial source might look differently, as can be seen in a graph from Kardashev's paper that shows how CTA-102's and CTA-21's spectra differed from natural source Virgo A. By addressing the challenges of distinguishing between artificial and natural radio sources, Kardashev laid the groundwork for future CETI searches. His considerations regarding angular dimension, variability, and spectral patterns formed the basis for subsequent studies and discussions in the field, shaping the methodologies and approaches employed in CETI in both the United States and the Soviet Union.

Kardashev's ideas were clearly influenced by communism and cosmism; his thinking on the nature of extraterrestrial intelligence was highly "Soviet" in character. With its centralized, technocratic, and expansionist vision, which mirrored the Soviet Union's pursuit of global influence, Kardashev's concept of supercivilizations reflected a projection of Soviet ideals onto the cosmos. Critically, Kardashev believed that this cosmic projection was inevitable. In the 1980s he wrote a paper titled "On the Inevitability of Super Civilizations," in which he laid out the likelihood that various types of civilization existed in the galaxy, ultimately concluding that all of them would eventually become united under one "higher" civilization. This ideological backdrop, prevalent during the Cold War period, coincided with the Soviet Union's ambitions to expand its influence over its satellite states and the unaligned third world. Kardashev's conceptualization of a future dominated by technocratic, expansionist civilizations that would unify lesser ones reflects a vision reminiscent of the Soviet approach. Kardashev's CETI pursuits intersected with the ideological landscape of the time, showcasing the complex interplay between science, politics, and philosophy.

In May 1964 the Soviets held their first CETI conference, the All-Union Conference on Extraterrestrial Civilizations, at the Byurakan Astrophysical Observatory in Armenia. It was at this conference that Kardashev first presented the ideas from his paper – along with

---

version reads that CTA 21 and 102 "display angular [dimensions] not less than 20 [seconds of arc]," which reverses the intended meaning. Clearly language barriers were yet another inhibitor of Soviet and American communication, and historians need to be alert to these issues when reading texts in either language.

other Soviet CETI pioneers, including his professor, I. S. Shklovsky, V. S. Troitsky, and V. A. Kotel'nikov. Following the success of the conference, the Soviet Council on Radio Astronomy of the Soviet Union Academy of Sciences created a section called "Search for Extraterrestrial Civilizations," which signaled state support for the newly established field of research.[53] Obtaining state support was crucial for the survival of the field, and to achieve it the astronomers had to establish a link between Soviet ideology and CETI. In writing on the organizational features of Soviet science, historian Loren Graham notes that "the first characteristic of science in the Soviet Union . . . was the uncommonly large role played by the central government."[54] Crucially, Soviet scientists were compelled to align their work with Marxist dialectical materialism if they were to gain governmental support.[55]

The debate on extraterrestrial life in Soviet ideology slightly pre-dated the development of CETI in the Soviet Union. In the 1950s, scientists at the Soviet Institute of Astrobotany argued that, according to the principles of dialectical materialism, extraterrestrial life must exist, and not finding evidence of it on nearby planets such as Mars or Venus would be a "clear disproof of the philosophical basis of communism."[56] Such a bold claim demonstrated the importance of establishing compatibility between scientific disciplines and ideology in the Soviet Union during the mid-twentieth century. Given what was at stake, the astronomers at the 1964 Soviet CETI conference penned a document titled "Resolution of the All-Union Conference on Extraterrestrial Civilizations."[57] The document was published after the conclusion of the conference and stated that CETI was of "enormous scientific and philosophical significance," and that Marxist "materialistic philosophy has firmly rejected the concept of anthropocentrism," ergo supporting the notion that extraterrestrial intelligence may exist.[58] Finding proof of extraterrestrial life would be further confirmation that the Marxist worldview was the correct one. After having successfully established both the need for further study and the subject's adherence to state ideology, Kardashev became vice-chairman of the newly established section and was tasked with developing a formal research program on the problem of the search for and communication with extraterrestrial intelligence in the Soviet Union. He was also tasked with launching various scientific searches.

Before the All-Union conference, astronomers at the Pulkovo Observatory in Leningrad had decided to investigate Kardashev's hypothesis regarding CTA-21 and CTA-102. They presented their results at the conference.[59] The primary investigator, Yuri Pariiskii (another former student of Shklovsky's), explained that the team's observations showed the "expected dependence for artificial radio spectra on account of the linear variation of the quantum fluctuation power with frequency," which seemingly lent credence to Kardashev's suggestion that it might evidence of extraterrestrial intelligence.[60] Pariiskii pointed out, however, that in the United States Ken Kellermann had recently interpreted the spectra as falling "within the framework of the synchrotron radiation mechanism"; he was inclined to agree with Kellermann, since the agreement between the calculated and the observed spectra was "astonishing."[61] Put simply, if a credible natural explanation was available, Pariiskii believed that this was a conclusion preferable to making a claim of detecting evidence of extraterrestrial intelligence.

Even in the first few years of CETI, distinguishing between natural and artificial signals was of the utmost importance. Pariiskii's consultation of Kellermann's work and Kardashev's dependence on the CalTech catalogue underline how important international communication and information sharing were in radio astronomy and in the development of CETI. Pariiskii's observations may well have been the end of the CETI fascination with CTA-102, but in the following year Sholomitskii was asked by Kardashev and Shklovsky to conduct a further observation of CTA-102. His subsequent discovery of radio variability would set off a chain of events that gave CETI international attention and brought embarrassment to Soviet science.

A Supercivilization Is Discovered

Sholomitskii's telegram was published in February 1965, less than a year after the Soviet CETI conference was held, and, as we know, faced incredulity from the western scientific community.[d] Shortly afterwards, during

---

[d] At this point in the book I must pause to discuss primary sources. Researching Soviet history can present great challenges, which are due to a lack of archival culture and the existence of state censorship. In piecing together the events of the CTA-102 affair, I relied on oral history interviews with the parties involved, newspaper articles from both the Soviet Union and the United

Figure 3.3 An article published on April 13, 1965 in the *Coventry Evening Telegraph* shows the damaging effect the CTA-102 affair had on the global reputation of Soviet science. Bernard Lovell is quoted saying: "It is rather sad. The Russians are in some ways given to extravagant interpretations of their results."
Credit: Rebecca Charbonneau.

a seminar at the Institute, Kardashev – according to the recollections of his colleagues at Sternberg – made a remark about CTA-102 possibly being the product of an extraterrestrial civilization. Lev Gindilis, a friend and colleague of Kardashev's at Sternberg, recounts that a Telegraph Agency of the Soviet Union (TASS) reporter named Alexander Midler attended the colloquium and overheard Kardashev's remarks.[62] TASS was the central state-run news agency that provided news collection for both national news outlets (such as *Pravda*, *Izvestiya*, and *CT USSR*) and international news outlets that wished to report on Soviet subjects.

States or Britain, telegrams from TASS Archives Siberia, and an essay with the recollections of a TASS journalist published in Russia in 2019. As is often the case when one uses documentation that relies on memory, some accounts are mildly contradictory. Yet, as I will show, even these contradictions help us understand the nature of scientific communication during this period.

On April 12, 1962, TASS issued a telegram regarding the discovery of an artificial extraterrestrial signal by Soviet astronomers (Figure 3.3).[63] The telegram was headlined: "Another civilization is signaling us."[64] It began with the stunning claim: "Radio signals, detected from an object in space, might belong to the technology of a highly developed extraterrestrial civilization, declare Moscow astronomers." The telegram, which is broken into four parts, then went on to explain Kardashev's hypothesis from his "Communications" paper. Midler paid particular attention to the size of CTA-102, quoting Kardashev as having said that "if this source is not really created by nature, but is indeed the creation of reasonable creatures, then it should be very small in size."[65] He then noted that western astronomers, most notably those at Jodrell Bank in England, had recently measured the size of CTA-102, and reported that it was indeed "extremely small."[66] In the third part of the telegram, Midler explained the significance of discovering radio variability, claiming that, "until Sholomitskii, no one anywhere had detected a source of radio emission in space which weakened and then strengthened, like a distant lighthouse", using the lighthouse imagery to further suggest an artificial origin.[67] The discovery of variability alone was "an enormous discovery," argued Midler.[68] Because of the high stakes in such a groundbreaking discovery, Midler made sure to emphasize that the astronomers had heavily vetted their results and that "Shklovsky and employees of the laboratory of the Sternberg Astronomical Institute [had] been testing ways to 'disprove'" the data. In the fourth part, the telegram took a radical shift. It began by stating: "Now, the scientists have no doubt. They say – 'We have a matter which may be one of the most outstanding discoveries in the history of radio astronomy.'"[69] The final part of the telegram also included a measured statement from Shklovsky, in which he expressed excitement for a new discovery but cautioned that there was still work to do in order to determine what the cause of the signal was. The telegram concluded with a claim to counter Shklovsky's restrained response: "Dr Nikolai Kardashev holds a more defined opinion: a supercivilization is discovered."[70]

* * *

Since the telegram confusingly conflated radio variability with extraterrestrial intelligence, it naturally garnered much attention from the media. The following day there was a large press conference held at Sternberg.

The massive response from the media and the public appeared to have caught the scientists of the institute off guard. In his autobiography, Shklovsky wrote that, on that day,

> [the] entire courtyard was crammed with luxurious foreign cars belonging to some 150 of the leading accredited correspondents in Moscow. I led off with a few conservative and skeptical principles. Sholomitskii was extremely restrained.[71]

Indeed, in recalling the press conference in his autobiography, Shklovsky represented himself as being fairly dismissive of Kardashev's claims, even going as far as to say that Kardashev was filled with "adolescent optimism" if he believed that the variability in CTA-102 was of intelligent origin.[72] At the press conference Shklovsky also derided TASS for its hasty publication, stating that "it is our right to ask journalists that they respect their great responsibilities, which does not always happen."[73] But, despite attempts by Shklovsky and Sholomitskii to undermine the extraterrestrial hypothesis, the narrative quickly spread around the world. Soon after the alleged discovery was reported, Kardashev received a telegram from Frank Drake, congratulating Sternberg and asking for more details – one of the earliest instances of direct communication between Soviet and American astronomers on the topic of CETI.[74] In his own autobiographical writings, Drake recalled how Shklovsky later told him that Sternberg had received a second telegram from America, one from Caltech scientists, who informed them that the signal was definitely not a signal from extraterrestrial intelligence. This telegram stated that the source had recently been identified by Caltech scientists as a quasar.[75] Unfortunately, as required by "protocol," the results of that identification could not be made public until they were published in a scientific journal. The dissemination of results was often delayed in the Soviet Union, which was partly the work of the Soviet Foreign Censorship Committee.[76]

International news media and popular culture nonetheless ran wild with the claim that the Soviets had made first contact with extraterrestrials. The 1960s American rock band The Byrds even wrote a song titled "C.T.A 102," which included the following CETI-themed lyrics:[e]

---

[e] I will warn you, before you go look it up, it is not a very pleasant song to listen to.

> C.T.A. 102, year over year receiving you
> Signals tell us that you're there
> We can hear them loud and clear
> We just want to let you know
> That we're ready for to go
> > Out into the universe
> We don't care who's been there first
> > On a radio telescope
> Science tells us that there's hope
> Life on other planets might exist.[77]

Although the detection of extraterrestrial intelligence was null and the event was a débacle for the reputation of Soviet science, it ultimately raised the public profile of CETI. This was not necessarily a positive thing.

On April 13, 1965, a newspaper in the United Kingdom interviewed Sir Bernard Lovell about the alleged discovery. In the article, Lovell is quoted stating that the telegram's report was "rather sad," then going on to say that "Russians are in some ways given to extravagant interpretations of their results."[78] Lovell's public condemnation would have had a significant impact in the Soviet Union, as he was still recognized there as an elite and renowned member of the international scientific community. Journalists in other countries were similarly patronizing. In the *New York Times*, an article titled "Natural Origin Indicated" noted that Shklovsky was a "long-standing enthusiast" of the idea of extraterrestrial intelligence, and pointed out that, just two months earlier, American astronomers at CalTech had identified CTA-102 as a quasar.[79] In fact CalTech had even been able to obtain a spectrum of CTA-102 and measured its redshift, but, given barriers to communication, this information had not yet reached the Soviet Union either.[80] Newspapers also pointed out that Sholomitskii's observations had been conducted at an undisclosed facility (he could not name the Yevpatoria array in his announcement on account of its military affiliation), and this cast further disbelief on his results. In another *New York Times* article, published the week after the Sternberg press conference, Walter Sullivan, a renowned science writer, pointed out this "curious contradiction" in Soviet and American astronomy:

The Americans believe that the object ... is so distant that it and other such objects may open the way to a deeper understanding of the very nature of the universe. The Russians say it is comparatively close and may be trying to attract our attention.[81]

Why this significant difference in opinion? Sholomitskii believed that, given the strange distribution of its spectrum, CTA-102 must lie "in or near the Milky Way galaxy," because an object that was very large and far away, like a distant galaxy, could not possibly pulsate in the 100-day rhythm he had observed.[82] Such an occurrence, Sullivan noted, would be like watching "an elephant can do the twist."[83] American astronomers, on the other hand, believed the object to be incredibly large and distant, but agreed that the 100-day pulsation was strange, and therefore were inclined to disregard Sholomitskii's results, especially given the secrecy over the Yevpatoria facilities. If Sholomitskii was right, Sullivan added, "there must be some way, so to speak, to get the elephant to do the twist."[84] Kellermann believed that the variability detected by Sholomitskii was "'theoretically impossible' since light travel time arguments meant that the source would need to be so small, that any radio emission would be self-absorbed."[85] Kellermann later admitted that "we now know that the radio emission from CTA-102 does vary at 30 cm on the time scales reported by Sholomitskii, as do many other quasars, and that this phenomena [sic] is now understood to occur as a combination of relativistic beaming and interstellar scattering."[86] Nevertheless, the western acknowledgment of Sholomitskii's discovery came too late; the combination of his covert observations using the Yevpatoria array and the claims of extraterrestrial intelligence contact all but discredited the discovery at the time.

Soviet science was made to look foolish by the international scientific community and by the media. At this point, it becomes important to defer to the journalist's perspective. Midler wrote an account of his memory of the CTA-102 events for a Russian book celebrating Shklovsky's life published in 2019, the contents of which highlight yet another dimension of mixed signals in Cold War communication. Aptly titled "! Или ?" – "! Or ?" – Midler's essay explained how a punctuation mishap led to this international scandal around Soviet science. Rather than "overhearing" a conversation, as Gindilis recounted, Midler claimed to have interviewed Shklovsky and Kardashev after a Sternberg lecture during which Kardashev made remarks about CTA-102 possibly being of intelligent origin. Importantly, this conversation happened on April 12, the anniversary of the day when Yuri Gagarin became the first human being to journey into space – a day formally celebrated as Cosmonautics Day in the Soviet Union.

Anniversaries were of particular importance in the Soviet Union, and so Midler and his team hurried to get the telegram that announced this discovery out as quickly as possible. In his retrospective, Midler said that he formulated his headline about the discovery as a question: "An Extraterrestrial Civilization Is Signaling Us?"[f] He claimed that the question mark was removed during the editorial and publishing stages, which turned his question into an assertion. After Shklovsky made disparaging remarks about TASS in the subsequent press conference, Midler approached him, distressed, saying: "What have you done! I will be fired, and three levels of editors and bosses will also be fired. I have only just started in this job and now Reuters has published an article titled 'Soviet Scientists Refute Soviet Telegraph Agency.'"[87] This negative international attention had dire implications for TASS and for Midler. Other sources, including Gurvits and Gindilis, suggested that Midler was even frightened for his life.[88]

Shklovsky, who as we heard previously had also experienced threats on his life from the Soviet bureaucracy, allegedly told Midler that he would take care of the problem and "write an article in *Pravda* in [Midler's] defense."[89] The article, published on the front page, was titled "New in Radio Astronomy: A *Pravda* Interview" and was framed as an interview with Shklovsky. But, unlike in other *Pravda* interviews, no interviewer was listed as author and there was no back-and-forth questioning. It seems likely that the entirety of the article was written by Shklovsky, as an editorial of sorts that was posing as an interview.

It is not surprising that Shklovsky would be granted this much control

---

[f] In the Russian language, the absence of punctuation characters such as the question mark or the exclamation mark can make it challenging to determine whether a sentence is a statement or a question. This feature is primarily due to Russian's flexible word order and reliance on intonation to convey meaning. Russian uses a subject-verb-object (SVO) word order as the default, but allows for flexibility and variation in sentence structure. This means that word order alone is not always sufficient to determine the intended meaning of a a sentence.

of the narrative in a state newspaper. In the mid-1960s he was considered an eminent and respected scientist in the Soviet Union, having previously won a prestigious Lenin Prize for his role in the Space Race.[90] Shklovsky was therefore given more leeway than others might have been. In the "interview," Shklovsky began by giving context for the discovery, briefly explaining how the discovery of variability was made, and praising Soviet technology, which helped make the discovery possible. Despite his qualms, expressed earlier, about the status of CTA-102 as evidence of extraterrestrial intelligence, in the *Pravda* article Shklovsky was not shy to mention Kardashev's hypothesis. He referred to the paper in which Kardashev predicted that CTA-102 could be artificial, saying that, "according to Kardashev's hypothesis, these sources could be radio signals of cosmic civilizations far from us."[91] Although he noted that there could be several different natural explanations for the discovery, he concluded with a remark that was titillating, yet conservative:

> Of course, one cannot exclude the exciting hypothesis that an artificial signal from an extraterrestrial civilization is being observed. New observations are necessary, however, for the hypothesis to become a scientific fact.[92]

In writing this editorial, Shklovsky was able to downplay the initial claim that CTA-102 was assuredly evidence of extraterrestrial intelligence, but still allowed for the possible veracity of Midler's claims. He also argued that CETI was a viable outlet for scientific investigation. As Midler later put it, Shklovsky wrote "as if there were no guilty persons" in the whole débacle; both the enthusiasm for the possibility of extraterrestrial civilizations and the exciting discovery of natural radio variability were valid responses.[93] Perhaps thanks to Shklovsky's article (although this cannot be definitively proved), Midler was neither fired nor imprisoned, and CETI research in the Soviet Union continued to thrive.

4ᵃ

First Contact
The Relationship between Carl Sagan and I. S. Shklovsky

Working as a C/SETI historian is a bit of an unusual way to put dinner on the table. Understandably, I am often asked how I first became interested in this career.

When I was a history student at Oxford, I was tasked with choosing a topic for my master's dissertation. As a young historian of science in training, I knew I was interested in space history, but I was not sure where I could make a new contribution. The history of the space race had been written to death, right? The rhetoric of competition, "the right stuff," and the hostilities between communism and capitalism were already fully entrenched narratives in that history. Still, determined to enter the scholarly conversation, I spent one afternoon perusing the shelves of the History Faculty Library in the Bodleian, hoping for inspiration to strike.

One dusty book on the shelf caught my eye – a book by Carl Sagan that I did not recognize. As an avid reader of Sagan's works, I thought I had read everything he had ever written, but this book, *Intelligent Life in the Universe*, I had never heard of. Upon further investigation, I realized that this was because it had long been out of print: the book was published in 1966, well before Sagan became a household name, and the last edition was printed in 1998, not long after I was born. No wonder I hadn't seen it before.

Delighted to have a new book by Sagan to read, I took the copy home. Once I began reading it, however, I realized what a strange book it was. For one, it had a co-author, someone I had not heard of. His name was

---

[a] Parts of this chapter have been reworked from Charbonneau, Rebecca. "Historical Perspectives: How the Search for Technosignatures Grew out of the Cold War." In Berea, Anamaria, ed., *Technosignatures for Detecting Intelligent Life in Our Universe*. Hobocken, NJ: Wiley & Sons Inc., 2022.

Iosif Shklovsky (a name that is hopefully familiar to *you* at this point) – a Soviet astrophysicist.

Second, the book was not written the way co-authored books typically are – in one unified text – but looked as though the authors had taken turns writing, almost in a clunky dialogue. Sagan's and Shklovsky's contributions were delineated from each other, marked by small symbols designed to indicate who wrote each paragraph. It made the book a little awkward to read. And to top it all, the subject was quite peculiar. The book began very much like a science textbook, explaining basic concepts in astrophysics. But, as it went on, it covered the burgeoning ideas in the search for and communication with extraterrestrial intelligence. It made unusual suggestions, pondering whether the Earth could have been visited by alien civilizations in our own civilization's ancient past. It had newspaper cartoon strips that marked the beginning of many of the chapters.

An odd book, to say the least. But beyond the suppositions of ancient astronauts or alien artifacts, I was most curious about the authors. Carl Sagan was a young man in his twenties when he wrote it; this was not yet the popular science superstar we now know and love, but rather a plucky postdoc who spent some time advising NASA as a side gig (Figure 4.1).[b] Shklovsky, on the other hand, was an established and respected scientist in the Soviet Union; by 1966 he had already won awards for his role in the Soviet space program (Figure 4.2). The early 1960s, when the book was being written, were some of the most tense years in the Cold War and the Space Race. That decade was marked by moments when the world found itself on the brink of nuclear war, and also by intense competition between the Soviet Union and the United States in the realm of outer space. How strange, then, to find a collaborative book by a Soviet and an American, both of whom were involved in their nations' respective space efforts – and a book about aliens, of all things.

This chapter is a history of that book. A book about a book may not sound terribly appealing, but I assure you that this is no ordinary tome. In fact it is much more than a book. In many ways, *Intelligent Life in the Universe* was the embodiment of a relationship, of a moment in time, of

---

[b] OK, OK, it's a little more nuanced than that. But stay with me for brevity's sake. Trust me, I was *also* a plucky postdoc who spent her late twenties writing about aliens.

Figure 4.1 Carl Sagan, the renowned astrophysicist and science communicator, stands proudly before a mockup of NASA's Viking Lander in Death Valley, California, in 1980, nearly 20 years after he first contacted Shklovsky in 1962. Sagan was a key figure in NASA's Viking program, which successfully landed the first spacecraft on Mars in 1976 with the intention of searching for biological life on the red planet. Sagan dedicated much of his life to the search for life in the universe.
Credit: "COSMOS A PERSONAL VOYAGE" / Druyan-Sagan Associates, Inc.

Figure 4.2 Iosif S. Shklovsky was a Soviet astrophysicist. For most of his career he served as a professor at Moscow University and played a pivotal role as the founding head of the radio astronomy department at the Sternberg Institute. He also held the position of chief of the astrophysics department at the Institute of Space Research in Moscow from 1972 to 1985.
Credit: https://phys-astro.sonoma.edu/brucemedalists/iosif-shklovskii.

the birth of a science and a hope. It is the book that first got me interested in CETI. If you read *Intelligent Life*, you will learn a lot about space and astronomy. But if you read about how and why *Intelligent Life* was written, you will learn a lot about our own world.

But, before we can dive into *Intelligent Life* and the relationship between Carl Sagan and Iosif Shklovsky, I must first address an embarrassing omission in the present book. I have thus far shown how the intervention of technology revolutionized the approach to the question of the plurality of worlds. Radar technology facilitated the growth of radio astronomy as a postwar discipline. Radio astronomy and its infrastructure developed rapidly in countries that had played a role in World War II radar development, including Britain, Australia, and the Netherlands. The United States had a late start in developing radio astronomy and was largely motivated by its military applications and the scientific–technical competition with the Soviet Union. I have suggested that it was the development of these facilities that prompted the development of CETI.

But this suggestion leads naturally to a pressing question, one yet left unaddressed: why did the other countries with thriving radio astronomy programs – I mean, other than the United States and the Soviet Union – *not* develop an interest in CETI? If radio astronomy infrastructure led to the development of CETI, why don't we see robust CETI programs in countries such as Britain in the 1960s? Why was it largely restricted to the United States and the Soviet Union? Why was *Intelligent Life in the Universe*, arguably the first book on CETI, authored by a Soviet and an American?

This is where the Cold War characteristics of CETI become especially significant in our pursuit to understand the field. It was not simply the development of technology, or the progression of scientific ideas, that led to CETI. It was distinctly the Cold War environment that produced the field as we know it. And this insight helps us understand why CETI developed nearly exclusively in the United States and Soviet Union in the mid-twentieth century, although radio astronomy infrastructure existed elsewhere too. The Space Race naturally created a popular space culture in the United States and in the Soviet Union, and the two respective governments had an interest in fostering public support. A "space age," so to speak, promoted imaginative thinking in the realm of space studies, including CETI. Further, though it may seem contradictory, the hostility between the United States and the Soviet Union encouraged the development of CETI. CETI researchers faced great challenges in communication and collaboration across the Iron Curtain. There was a true the crisis of communication with the "alien" – both on other planets and on Earth. CETI fostered a global perspective that prompted CETI pioneers to attempt to shed their nationalism as they cooperated with international peers, while also creating a new cosmic framework within which to observe the global conflict and terrors of the Cold War.

The Power of Friendship

If you were a child growing up in the 1980s or 1990s, as I did, you may be familiar with the television show *My Little Pony*, which follows the adventures of a group of colorful, anthropomorphized ponies. The overall theme of the franchise was the power of friendship (in fact the show had a spin-off in the 2010s that was titled *Friendship Is Magic*). We may be inclined to regard such a message childish, but there are strong truths underlying the schmalzy parables that we find in twentieth-century children's programming. In history we do not explicitly refer to the power of friendship per se, but we do have a similar concept: citizen diplomacy. Citizen diplomacy refers to the power individual citizens hold on the larger goals of their nation. When private citizens such as the scientists in the United States and in the Soviet Union pursued relationships with one another, this sometimes had a cascading effect on the relationships between the two countries. Citizen diplomacy posits that everyday people can make a big impact on global relations, be it through activism, through charity, and, yes, even through the power of friendship. But during the Cold War these private friendships were often troubled by the competing goals of the larger nations.

On the American side, the idea that science was greater than politics was key in motivating American scientists to cooperate across borders in their capacity as citizens. As historian of science Audra Wolfe has noted, there were individual and cultural dimensions of warfare that were practiced by both the United States and the Soviet Union during the Cold War, and this affected how scientific ideology was portrayed during this period. Wolfe argues:

The US foreign policy establishment saw a particular way of thinking about scientific freedom as essential to winning the global Cold War – and not just because science created weaponry. Throughout this period, the engines of US propaganda amplified, circulated, and, in some cases, produced a vision of science, American style, that highlighted scientists' independence from outside interference and government control. Science, in this view, was apolitical.[1]

The idea of using apolitical science as a Cold War strategy led to the creation of scientific exchange programs between the United States and the Soviet Union. The development of scientific exchanges between the United States and the Soviet Union dates back to the 1950s and to President Eisenhower's assertion that individuals should "leap governments" or "if necessary evade governments," because to do so would allow people from different nations to learn from one another and grow, which in theory could lead to more peaceful interactions.[2] Yet beyond science alone there were multiple elements to these exchanges. Gerson Sher, former Program Coordinator for US–Soviet and East European Programs at the National Science Foundation, has noted that even the word "exchange" "brought to mind a carefully calibrated and planned transaction, not unlike an exchange of spies."[3] Indeed, like many facets of Cold War scientific activities, scientific exchange programs had dual motivations: to promote the ideology of science as an apolitical and peace-generating entity, but also to obtain human intelligence on Soviet science and technology. As we saw with Lovell's reporting on Yevpatoria, gathering intelligence on Soviet technology was a fundamental factor that determined the US Cold War strategy.

Yet, as Wolfe notes, the scientists in the United States were "both building genuine friendships with their Soviet counterparts and collecting scientific intelligence"; these relationships were complex and multidimensional.[4] It was not necessarily one or the other – there might have been political or militaristic elements involved, but the friendships were also real. One astronomer I interviewed, Malcolm Longair from the University of Cambridge, explained that his motives for taking up an exchange to work with the Soviet astrophysicist Vitaly Ginzburg at the Lebedev Physical Institute in 1968 were largely curiosity and admiration: "I wanted to actually understand the Russian character, the Soviet character more because it was something else... and of course the music is absolutely out of this world."⁵ Longair was genuine and enthusiastic about his desire to cooperate with scientists in the Soviet Union and to learn from both their scientific theories and their culture. Yet Longair's exchange program through the Royal Society came with a stipulation: he had to complete a confidential report that detailed his visit to the Soviet Union, much like Lovell's 1963 diary and memorandum. In some ways, scientific exchange programs were an even more valuable approach to intelligence gathering than traditional human intelligence strategies with spies; after all, US and British scientists formed genuine connections and therefore sometimes received insider access and trust from their Soviet counterparts; thanks to them, their home governments could take an intimate look at Soviet science, technology, and institutions.

This governmental use of friendly scientists for intelligence gathering may well have had another impact: an increasing desire in some scientists, and especially CETI scientists, to cooperate *surreptitiously* with their Soviet peers. In 1960 Carl Sagan began a two-year postdoctoral fellowship at the University of California, Berkeley, where he was preoccupied with the problem of finding life in the universe. One day during his fellowship, he received a call from a US general who informed Sagan that he was presently escorting three Soviet scientists around Los Angeles and asked if he would like to join them. One of the Soviet scientists in LA was Alexander Alexandrovich Imshenetsky, who was, as Sagan described, "in charge of the Soviet effort for constructing instruments to search for extraterrestrial life."⁶ Sagan was invited to join the group, given Imshenetsky's specialization and its overlap with his own research. Sagan later wrote about this event in his semi-autobiographical collection of essays, which is tellingly titled *The Cosmic Connection: An Extraterrestrial Perspective*.⁷ He later described this meeting as the "first such contact" he would have with a Soviet scientist, and upon meeting Imshenetsky he eagerly engaged him in a conversation on extraterrestrial life.⁸

Also in attendance with the group was a man to whom, in his retrospective essay, Sagan referred by the false name "Igor Rogovin" – supposedly a translator from the Library of Congress. At one point during the meeting with the Soviets, Sagan was left alone with Rogovin, and was asked what he had "[found] out" during his conversation with Imshenetsky.⁹ In his essay, Sagan regrettably notes that at that young age he was "unwise in the ways of the world" and "politically unsophisticated," and so he enthusiastically summarized what he had learned from Imshenetsky, before realizing that Rogovin was not a translator, but an intelligence agent.[10] Sagan claimed to have berated Rogovin, scolding him that "it was possible to have a conversation with a Soviet scientist that was intended for the benefit of science, rather than for the benefit of American military intelligence services."[11]

Sagan's subtitle *An Extraterrestrial Perspective* is a clear sign that in 1960 the young scientist was already making an association between scientific internationalism and the pursuit of extraterrestrial life. After his hostile interaction with Rogovin, Sagan contacted the CIA to lodge a complaint about the conduct of the agent, only to discover that the CIA did not think that Rogovin was one of their agents. Sagan was asked for discretion, as the CIA was worried about such a story reaching the public, "particularly after the 'bad press' they had been getting about the Bay of Pigs."[12] After two weeks of investigation, it was revealed that Rogovin was an agent who worked for the Air Force Intelligence, not the CIA.

The experience had an impact on the young Sagan. He was surprised that "Soviet plans for the search for life elsewhere . . . could be considered of interest to Air Force Intelligence," though of course we now know there was a strong connection between the two.[13] Sagan was also disquieted that it "had taken about two weeks for the Central Intelligence Agency to determine the employment of a member of a fellow US intelligence operation."[14] At the time, the naïve Sagan was appalled that intelligence agents would try to use an "innocent young [scientist]," as he described himself, to gather intel on Soviets.[15] He believed that their efforts aimed to "detract from the credibility of legitimate scientific exchanges among scientists in different countries," even though in his view such exchanges were "particularly necessary in an age that hangs a thread away from nuclear destruction, and in which scientists have access to at least half an ear of the politicians in power."[16]

This episode in Sagan's life is significant to the history of CETI. It formed the basis of his later internationalist philosophy, which would lead to the cosmic mirror perspective he developed in projects designed to message extraterrestrial intelligence. It helped him to start by recognizing the presence and mission of US intelligence agencies, an understanding that would later aid him in navigating the difficult distinction between artificial signals of extraterrestrial and of terrestrial origin. And, finally, his frustration with political interference in science energized him to form friendships with Soviet colleagues, something that would later shape the character of US and Soviet CETI.

A Most Important Find

The advent of radio astronomy shifted the investigation of communicating with extraterrestrial intelligence from the speculative to the truly technical: it gave scientists the opportunity to test their theories by making strategic radio observations of the cosmos. Although experiments began as early as 1931, with Karl Jansky's discovery of the cosmic sources of radio waves,[17] the formal discipline of radio astronomy was a direct product of World War II. After the war, former wartime radio engineers had a great interest in using the recently developed radar and radio communication technologies for scientific purposes, and scientists in the nations that had been involved in the war began to pursue research in the newly burgeoning field of radio astronomy.[18] Yet it was not until the discovery of the 21-centimeter hydrogen line – "the magic frequency," as historian Steve Dick has called it[19] – that CETI began its development.

The prediction of the 21 cm hydrogen line is generally attributed to Hendrik van de Hulst, a Dutch astronomer and mathematician who in 1945 published a paper that suggested that the transition of neutral hydrogen at 1420 MHz should be theoretically observable using radio telescopes.[20] This was a highly significant insight as far as the development of radio astronomy goes, because of the abundance of hydrogen in the universe. Just a few years later, in 1951, Harvard University astronomers Harold Ewen and Edward M. Purcell were the first to observe the line. Since then, the hydrogen line has become a fundamental part of observational radio astronomy, allowing astronomers to map the structure of the Milky Way and other galaxies as well as the large-scale structure of the universe, since hydrogen is the most common element in galactic formations. So significant to radio astronomy was its discovery that it has even been memorialized in song, through a chorus that reminds listeners that

> Ewen and Purcell caught the radiation line
> Of interstellar hydrogen, a most important find.[21]

When examining the history of the discovery of the line, however, another name besides van de Hulst's sometimes appears: Iosif Samuelovich Shklovsky. Several sources, including *The Biographical Encyclopaedia of Astronomers* (2007) and Frank Drake's *Is Anyone Out There?* (1992), claim that Shklovsky predicted the existence of the 21 cm hydrogen line entirely independently of van de Hulst. This is a mischaracterization of what occurred, but a somewhat understandable one if you are unfamiliar with Soviet journal publishing. Astronomer and historian of astronomy Woodruff Sullivan, who conducted interviews with Shklovsky before his death in 1985, noted in *Cosmic Noise* that Shklovsky was not able to obtain van de Hulst's paper in the Soviet Union because of the publication issues mentioned earlier. Instead, Shklovsky found a brief note about van de Hulst's discovery in an astronomical review paper published in 1947. The mention of this prediction in the review "lit [him] on fire" and inspired him to attempt to calculate the transition of the line on his own, without access to van de Hulst's original paper.[22] In 1949, the same year in which he received his doctorate, Shklovsky published a paper on the possibility of observing monochromatic emissions from the galaxy.[23] Shklovsky credits van de Hulst in the first line of his paper, asserting: "Van de Hulst was the first to point out the probability of the existence of monochromatic radio emission from the galaxy."[24]

Nonetheless, Shklovsky's semi-independent calculations of the 21 cm hydrogen line helped establish him as a significant early contributor to the development of radio astronomy in the Soviet Union. But there is another way in which the 21 cm line had an impact on Shklovsky's life. Once observed by Ewen and Purcell, the newly important status of this line to radio astronomy inspired two physicists, Philip Morrison and Giuseppe Cocconi, to propose using it in searches for extraterrestrial intelligence. Published in *Nature* in 1959, Morrison and Cocconi's paper advocated a search for artificial signals on the hydrogen line; thus it became the first scientific publication to propose a CETI observational technique.[25] The authors' rationale for observing at 1,420 MHz was that, if an extraterrestrial civilization wanted to send a signal at a frequency that Earth was sure to detect, it would make most sense for them to broadcast it at a frequency that is important to radio astronomy, where humans are already looking. Shklovsky was "greatly impressed" by Cocconi and Morrison's article, which added such great potential purpose to the line he had played a role in discovering.[26] In April 1960, shortly after the publication of the Morrison and Cocconi paper, Drake conducted Project Ozma, searching for signals near the hydrogen line. Although his results were non-conclusive, Project Ozma excited the scientific community, including Shklovsky. Shortly after learning of Project Ozma, Shklovsky published his first CETI paper in the Russian journal *Priroda* [*Nature*].[27]

The paper began with an acknowledgement that "the very title of this paper will seem fantastic to readers of *Priroda* . . . is it even possible to discuss such an unusual problem on the pages of this serious journal?" He then broke down the problem into sections, bringing in other areas of astronomical inquiry, for example by asking, "are there other planetary systems?" (something that was not yet definitively known in the 1960s).[28] He concluded the paper by referencing to Morrison and Cocconi and Project Ozma, noting their "elegant idea" of using the 21-centimeter hydrogen line for communication, since if extraterrestrial intelligence consists of "reasonable beings at a high level of technological development," they too "must conduct intensive studies of the Cosmos precisely on this wavelength."[29] In other words, this is where they will be looking and where they would assume us to be looking as well. Importantly, Shklovsky's paper essentially made the argument, for the readers of *Priroda*, that CETI is not just *a* significant field in astronomy but *the* field. He argued that all areas of astronomical inquiry support and lead up to asking the question: is there intelligent life in the universe?

Universe, Life, Mind

The start of the Space Race deepened Shklovsky's fascination with the possibility of searching for extraterrestrial intelligence. In January 1959, shortly before he learned of the Morrison and Cocconi paper, the Soviet mission Luna 1 was launched. After the glowing success of the first three Sputniks, Luna 1 became yet another important achievement for the Soviet Union at the start of the Space Race: it was the first spacecraft to reach the Moon.[30] There was a small dilemma, however: in 1959 the Soviet Union did not yet have a radio telescope capable of tracking its satellite and probe launches. It is for this reason that Sir Bernard Lovell became a figure of much acclaim in the Soviet Union. His Mark I telescope became the first telescope in the West to track the launch of Sputnik I in 1957, confirming the Soviet achievement to the rest of the world.[31] Without their own radio telescope, obtaining the exact coordinates of their rocket launches was a real challenge for the Soviets.

For the launch of Luna 1, Shklovsky had proposed a solution. With the support of the "chief designer" of the Soviet space program, Sergei Korolev, Shklovsky and his team at the Sternberg Astronomical Institute at Moscow University designed what they called an "artificial comet," to aid optical observations of rocket trajectories.[32] This "comet" was composed of a luminescent cloud of sodium gas that would be ejected from the rocket during launch and photographed, making the spacecraft visible at high altitudes and therefore easier to track optically.[33] For his work on the Luna 1 rocket telemetry, Shklovsky was awarded the Lenin Prize, the highest award for scientific achievements bestowed by the Soviet government.[34]

This success put him on good terms with the president of the Soviet Academy of Sciences, Mstislav Vsevolodovich Keldysh (who was sometimes referred to as the "chief theoretician" of the Soviet space program, in conjunction with Korolev). As a result of his success, Shklovsky was invited to group meetings, held in Keldysh's office, for "regular [discussions] of space projects."[35] During one such meeting in 1961, Keldysh reminded the group that the five-year anniversary of Sputnik I's launch was only one year away and "should be properly celebrated."[36] As previously mentioned, anniversaries held special significance in the Soviet Union, and this was especially true of five-year anniversaries, which evoked the five-year plans – the Soviet economic strategy for moving the state toward communism. Shklovsky, whose imagination had so been swept up by the Morrison and Cocconi paper and the romance of the burgeoning space age that he felt "like a kid who'd fallen in love," eagerly proposed that he write a popular science book exploring the idea of extraterrestrial life.[37] Keldysh approved, but this meant that there was only one year to write the book. What might have been a problem to some, Shklovsky saw as an opportunity.

That was because in the Soviet Union, in addition to the censorship of foreign scientific journal publications, there were also censorships placed on internal publications. In particular, it was a big challenge to get through the censors any work on "space" during the Space Race.[38] The main organization of press censorship was called the Main Administration for the Protection of Military and State Secrets in the Press under the Soviet Union Council of Ministers – for short, Glavlit – an acronym of the Russian name.[39] Glavlit had an associated office – the Commission for Research and Exploitation of Cosmic Space – which appointed specialized censors for space sciences. During the space age, "every book, article, radio or TV broadcast in any way connected with space flights [had to] have an authorization from [that] censorship office."[40] However, given the significance of publishing the book for the fifth anniversary of Sputnik's launch, Shklovsky suspected that he "would have a better chance of escaping the embraces of the general censorship," as censors would have less time to review it if it was published on time.[41]

Getting past the censorship was particularly important for a book on the subject of extraterrestrial life. Although by the 1960s Trofim Denisovich Lysenko's state-sanctioned views on biology had begun to fall out of favor, the official end to the ban on criticism of Lysenkoism was not lifted until the 1980s.[42] Lysenkoism refers to the pseudoscientific theories and agricultural practices promoted by the Soviet biologist Trofim Lysenko during the mid twentieth century. This doctrine rejected several aspects of classical genetics and embraced the Lamarckian concept of inheritance of acquired characteristics. Under Joseph Stalin's regime, Lysenkoism gained significant state support and was widely applied to Soviet agricultural practices, where it had disastrous results, including widespread famine. Scientists who objected to Lysenkoism faced serious consequences, for example job loss, imprisonment, and exile.

This was particularly relevant to Shklovsky's project: in writing about the development of life in the universe, he intended to "demolish" the theories of Alexander Oparin, a Soviet biochemist and "close confederate" of Lysenko who researched the origins of life on Earth.[43] Oparin first proposed the concept of chemical evolution, suggesting that life on Earth originated from non-living matter through a series of chemical reactions. He was a member of the Communist Party and an eminently respected scientist in the Soviet Union. Challenging Oparin was a potentially dangerous undertaking, not to mention challenging. After all, it was difficult to find reputable books on molecular biology in the Soviet Union during the Lysenkoist period.[44] Nonetheless, Shklovsky's bet paid off. His book, titled *Universe, Life, Mind* made its way past the censors in the rush to be printed in time for the Sputnik anniversary, and it was published in December 1963.[45]

Despite the initial challenges, *Universe, Life, Mind* became immensely popular in the Soviet Union and across the world. There have been seven Soviet and Russian editions of it to date, the most recent one in 2006. The book was promoted among the public because of its comprehensive and approachable exposition on modern astrophysics. The Soviet All-Union Society "Znanie" [Knowledge], an educational propaganda organization, awarded it first prize in the category "best popular science book."[46] It was well received by both popular and scientific audiences; but, interestingly, the book also sparked works in other disciplines, for instance philosophy. *Summa technologiae* (1964), a futurological treatise by the Polish philosopher and science fiction writer Stanislaw Lem, dedicated an entire chapter to a philosophical interpretation of *Universe, Life, Mind*. In his analysis, Lem spent most of his time reflecting on the longevity of cosmic civilizations and questioning whether the apparent lack of solid evidence for extraterrestrial civilizations means that the universe spawned only a "suicidal intelligence," which inevitably destroys itself.[47] This theme is significant because CETI works often produced reflection on existential issues – a key trait of the Cold War mentality.

Universe, Life, Mind's tremendous success was due in part to the large public impact of Soviet space activities. In the early 1960s the Soviet Union had dominated the United States in the "race to space," to use the label coined by President Kennedy in 1961.[48] By 1962 the Soviet Union had become the first nation in the world to launch a satellite into orbit. It put the first living creature in space, put the first human in orbit, took the first photograph of the far side of the Moon, launched the first space craft to reach the Moon, and sent the first probe to impact the Moon. Characteristically, the Soviet government capitalized on these victories and marked them with great displays of promotional propaganda in the form of films, posters, and parades. Iina Kohonen, an expert in space-related visual propaganda and photojournalism in the Soviet Union, has argued that the Soviet space program was a combination of "military aspirations, state propaganda and utopian thinking."[49]

It is the latter characteristic, utopian thinking, that largely fueled the Soviet public interest in space and led to the success of Shklovsky's book, which painted an exciting picture of the future of human cosmic exploration. Soviet depictions of the future in space presented a universe in which communism has succeeded, the world is peaceful, and humanity has dedicated itself to the pursuit of science and exploration of the cosmos.

As we have seen, this cosmist-like mentality was not limited exclusively to the Soviet Union. Americans similarly created narratives about a human destiny in space – albeit with a US-centric bent. While Soviet literature espoused spreading a communist ideology throughout the cosmos, Americans viewed space as a "final frontier," an extension of the driving force of manifest destiny.[50] Therefore, in the United States, public interest in space exploration was also growing by the time Shklovsky published his book. The United States had managed a rocky start in the competition; but, especially after Kennedy's "moon shot" exhortation, the US public imagination was captured by the space age. Yet the hopeful nationalist visions of a future in space were tainted by the hostility of the Cold War, which often used the same infrastructures of space exploration for military purposes.

Same Planet, Different Civilizations

While Shklovsky had been designing his artificial comet for the Soviet lunar program in 1959, Sagan was similarly involved in the race to space. At the age of only 25 years, he had been hired as a consultant to the newly established National Aeronautics and Space Administration (NASA).[51] By August 1962, just a few months before the publication of *Universe, Life, Mind*, Sagan had just seen the launch of NASA's Mariner 2 mission to Venus, which he had helped to design and manage. He was also a member of NASA's Planetary Biology Subcommittee and hoped to find signs of life on other planets through his work on NASA missions.[52]

In 1961 Sagan wrote a paper inspired by the Drake equation, which he had just learned about at the first CETI meeting in Green Bank. In it he attempted to calculate the number of intelligent extraterrestrial civilizations in the Milky Way galaxy. While he began the paper admitting that the "parameters" necessary to calculate the number "are poorly known," he nevertheless came to the conclusion that there are about a million "extant advanced technical civilizations in our Galaxy."[53] Perhaps having heard of the Soviet interest in CETI from Imshenetsky and feeling motivated to work with his colleagues across the Iron Curtain, Sagan sent a draft of this paper to Shklovsky on June 8, 1962,[54] inviting his comments.

Shklovsky responded positively and asked Sagan whether he might be permitted to incorporate the paper into his forthcoming book, which was set for publication at the end of the year. He ended his letter to Sagan with a joke, playfully pondering on the prospect of someday meeting in person. He acknowledged, however, the improbability of such an event, given the barriers between the United States and the Soviet Union, and compared it to the likelihood of an extraterrestrial visiting Earth.[55] Sagan agreed to have his paper incorporated into *Universe, Life, Mind*, and in return asked Shklovsky whether he had any plans to publish an English translation of the book in the United States. If not, Sagan proposed being allowed to aid with such a publication himself.[56] He cheerfully ended the letter suggesting that they meet at a conference soon to be held in Poland and mirrored Shklovsky's previous joke by adding that, if they were to meet, perhaps extraterrestrial contact was not that improbable after all![57] Alas, the older and more experienced Shklovsky was correct in his estimation of the barriers to scientific cooperation across the Iron Curtain. The two would end up not meeting for many years, until well after the publication of the English book. This was largely because of constraints put upon Shklovsky on account of his standing in the eyes of the Soviet authorities.

Although Shklovsky was held in high regard by his colleagues for his earlier work during the Space Race, his outspoken personality prevented him from enjoying the freedom of movement and access granted to some of his peers. His letter to Sagan regarding the unlikelihood of their meeting reflected the personal resentment Shklovsky held toward Soviet bureaucracy for limiting his ability to collaborate with international peers. In a short autobiographical account published posthumously, Shklovsky recounted his first expedition outside the Soviet Union: a research trip to Brazil in 1947, to observe a solar eclipse.

> I took it for granted that the forthcoming expedition to the Tropic of Capricorn, to a faraway Brazil as beautiful as anything in a fairy-tale, was just the beginning and that many more fine and soul-stirring things yet unknown lay ahead. After a poverty-stricken youth and the harsh suffering of the war years, the world had at last opened up for me.[58]

Unbeknownst to him at the time, Shklovsky would not be allowed to travel abroad for another 19 years. Despite his many accomplishments, including the Lenin Prize awarded him for his work during the Space Race, Shklovsky often faced travel bans and was never elected as a full member of the Soviet Academy of Sciences – which, he resentfully concluded, was because of his Jewish heritage and commitment to promoting human rights.[59] This was not atypical in the Soviet Union; another former Soviet astronomer whom I interviewed, Rustam Dagkesamanskii, the director of the Pushchino Radio Astronomy Observatory, told me with regret that he was often barred from travelling abroad to international conferences because he had refused to join the Communist Party:

> In the 60s, [I was invited] to the Communist Party ... but I said, "thank you, but I don't want [to]" ... If I would be [a] Communist Party member, I should say all things as the Community Party says. But I tried to keep my own opinions ... [if I had become a member of the Community Party] it would [have been] easier for me to travel [outside the Soviet Union].[60]

Soviet astronomers often faced restrictions to their scientific freedom if they did not align themselves with the correct ideologies or the correct political party, and in some cases these consequences could be severe.

In his memoir, humorously titled *Five Billion Vodka Bottles to the Moon* (1991),[c] Shklovsky shared his memories of the "astronomers purge" during Stalin's Great Terror of 1936–1938, when over two dozen leading Soviet astronomers were arrested, many of whom were later executed or died in the Gulag.[61] The purge was partly caused by an astronomy PhD student who failed his candidacy exam, and afterwards wrote a letter of denunciation of the astronomer who administered the examination, Boris Numerov.[62] Denunciations were common practice in the Soviet Union, especially during the Stalinist era. According to historian Sheila Fitzpatrick, denunciations largely fell into three categories: accusations of political disloyalty, "concealment of class identity," and "abuse of

---

[c] I highly recommend this book, which is filled with many amusing and sobering anecdotes on life as a Soviet scientist. In addition to dark and disturbing tales like those from the Astronomer's Purge, Shklovsky's characteristic humor manifests itself in many silly stories, such as a scene in which he puked in Stalin's childhood home, or one in which he was kicked out of a Parisian strip-tease joint.

power."⁶³ The state would investigate the accused and punish those whom it deemed guilty. This, however, created the problem of individuals who abused the system by using the state to "settle personal scores or advance the denouncer's individual interests," as was clearly the case with the failed PhD student.⁶⁴ In his denunciation, Boris Numerov was accused of having "foreign contacts," and shortly afterward the NKVD began investigating him. After he was arrested and tortured, he confessed "to being the organizer of a counterrevolutionary group of astronomers and geophysicists that had cooperated with German fascists and had engaged in wrecking, spying, and terror since 1929."⁶⁵ In his confession, Numerov listed nearly the entire astronomy community as co-conspirators, setting off a chain of subsequent arrests and denunciations that historian and foreign diplomat Robyn McCutcheon argued led to the "disappearance" of approximately "10 percent to 20 percent of all astronomers in the Soviet Union in 1935."⁶⁶ Only a graduate student at the time of the purge, Shklovsky avoided being arrested, but after his death in 1985 his account of the events that led up to the purge was published, and it gave a clear indication that the memory of the ordeal remained with him for the rest of his life.⁶⁷ Still, despite the threat of denunciation, Shklovsky often risked his life and freedom by standing up for what he believed in; and he continually criticized the Soviet Union for its anti-Semitism and restriction of scientific freedom (among other things).

Even as his fellow physicists occasionally disappeared or died under mysterious circumstances, Shklovsky would defend mistreated colleagues and condemn what he viewed as unethical transgressions committed by the Academy or by the Soviet government. In 1973 approximately 40 members of the Soviet Academy of Sciences signed a document condemning the Soviet nuclear physicist and political dissident Andrei Sakharov in a denunciation.⁶⁸ Shklovsky, on the other hand, sent in a strong letter of advocacy for Sakharov. As a result of this daring move, he was forbidden to appear at international scientific meetings he had previously been allowed to attend, such as those of the International Astronomical Union. When asked by western colleagues about Shklovsky's absence from international conferences, Soviet officials would reply that his health was too poor for him to travel abroad.⁶⁹ An American colleague encountered Shklovsky in the Soviet Union during this period of travel ban and inquired about his health, to which Shklovsky wryly replied:

"Yes, I have diabetes. Too much Sakharov."[70] This was a joke: the word for "sugar" in Russian is *sakhar*.

Although Shklovsky avoided arrest during the Stalinist era, his dissidence still put his life in danger sometimes. As we have learned, in his capacity as director of the Jodrell Bank Observatory, Sir Bernard Lovell went on an unprecedented scientific visit to the Yevpatoria Deep-Space Communication Centre in the Soviet Union in 1963, shortly after *Universe, Life, Mind* was published.[71] The official reason for the visit was a planned international scientific collaboration; the Jodrell Bank Observatory was to assist upcoming Soviet–US communication satellite experiments.[72] Lovell's refusal of the Soviet offer of residence in the Soviet Union as a member of the Academy, alongside his newfound insight into Soviet scientific infrastructure, may have led to his concerns (discussed earlier, pp. 68) that his Soviet colleagues attempted to brainwash him using radiation from a telescope's radar beam.[73]

It is unlikely that the events Lovell described truly occurred as he perceived them; those impressions were more likely the product of Lovell's paranoia and of the many myths and rumors that enveloped the secretive state. Nonetheless, his case is further evidence of the influence of Cold War anxiety. Lovell's memorandum also detailed an encounter with an unnamed Soviet scientist who implored him to invite Shklovsky to his laboratory in England, for Shklovsky's life was "in great danger."[74] The nature of the danger and the reason for it remained a mystery to Lovell, and there are no available sources that might explicitly offer a cause. I would argue, however, that the biology controversy stirred up by the publication of *Universe, Life, Mind* may have been behind these events.

I noted earlier that in *Universe, Life, Mind* Shklovsky intended to "demolish" the theories of Alexander Oparin, the Lysenko sympathizer. After the publication of the book, out of courtesy, Shklovsky sent Oparin a letter explaining his disagreement with Oparin's theories. Oparin allegedly shredded the letter, stuffed it back in an envelope, and returned it to Shklovsky.[75] Clearly tempers were running hot; during the Lysenkoist period many scientists who pushed back against the academic establishment went through life-threatening experiences. Why, then, did Shklovsky not ultimately suffer serious harm for his transgressions?

As Lovell noted in his diary, "Shklovsky later often appeared in the West . . . and as far as I knew he eventually died peacefully in the

Soviet Union."[76] Lovell also noted that he "did as requested" and invited Shklovsky to visit Jodrell Bank.[77] But Shklovsky was unable to make the visit, perhaps because his open dissent to the treatment of Jewish academics cost him the ability to travel outside the Soviet Union between 1947 and 1966.[78] Lovell appeared to suggest that perhaps his invitation held enough weight to aid Shklovsky, though he also admitted that he had "no idea."[79] I would argue, however, that Shklovsky's safety had less to do with Lovell's intervention than with the shifting scientific culture of the early 1960s. Historian of Soviet science Michael Gordin has noted, in a study that charts the downfall of Lysenkoism, that in 1962 (the year in which *Universe, Life, Mind* was published), Lysenko's theories and experiments began to come under heightened scrutiny in the Soviet Union. An Academy of Sciences commission tasked with visiting one of Lysenko's experimental farms was highly critical, noting that there was "widespread fabrication of data in order to cover up the shockingly poor, even ruinous, results. Lysenko, naturally, protested both the procedures and the findings of the commission."[80] A second report plainly noted:

> T. D. Lysenko confines himself only to general arguments and unsubstantiated assertions and presents unfounded accusations addressed to the commission, but he does not provide in this case any arguments, proofs, facts. Thus academician T. D. Lysenko is unable to refute even one of the commission's statements.[81]

Oparin had relied on Lysenko's strength as well as on his own close relationship with Stalin to maintain his scientific–political power. With Lysenko "finally ousted" and Stalin deceased, there was little that Oparin could do to punish Shklovsky for his daring publication.[82] This shift in the Soviet structure of scientific politics was important in the development of transnational CETI, for it allowed flexibility in the discussion of the origin of life in the universe. The increase in freedom allowed Shklovsky to share his formerly dissident perspectives and to make and maintain relationships with scientists around the world through extensive correspondence – including with Carl Sagan.

Intelligent Life in the Universe[83]

Yet establishing these international relationships through correspondence did not come easily, and the road to the publication of the English translation of Shklovsky's book was rife with difficulties. When Holden-Day, the US publisher, wrote to him in January 1963 to confirm the agreement to publish the translation with the aid of Sagan, it also expressed a commitment to expedite the publication process, aiming to release the book within a timeframe of three to four months.[84] More than eight months later, however, Shklovsky had no update from Sagan or the publisher. Frustrated by this silence despite the publisher's assurances, an irate Shklovsky wrote to Sagan, demanding information about the status of the American translation of the book.[85] Unbeknownst to him, however, Sagan had sent him several letters regarding the translation and its progress. Shklovsky had simply not received them – an example of yet another type of challenge that was common during the Cold War: the unreliability of postal correspondence across the Iron Curtain.

The issue of Soviet interference with the post is well documented. In the 1980s there was a concerted effort by the US government to address the deliberate interference with the flow of mail between the United States and the Soviet Union. A 1989 report on the history of mail interruption prepared for the US Committee on Post Office and Civil Service of the House of Representatives examined what they described as a "long existing problem."[86] The report claimed that, over many years, a "significant number" of items of communication sent from the United States to the Soviet Union had "disappeared, or were opened, inspected, and/or confiscated by officials of the Soviet Union, without the proper notification given to mailers."[87] Mail interruption was described as a "violation of human rights" by the authors of the report, who also presented data that indicated the rate at which letters were not delivered. In 1985, non-delivery of letters was at approximately 12 percent.[88] Given that 1985 was the year *perestroika* began, it is reasonable to assume that earlier decades had seen even higher rates of mail interruption, although data to support this hypothesis are difficult to obtain.

It would be naïve, of course, to assume that the problem of mail disruption existed exclusively on the side of the Soviets. As Edward Pessen, American historian, has pointed out, "what a nation's leaders call its policy is after all only its stated policy."⁸⁹ Although US policy was purported to be emblematic of a "free" society, there is also evidence of US interference with postal communication. As Pessen points out, "in blatant violation of the law creating it, the CIA kept files and spied on American citizens, tampered with and opened the mail of hundreds of thousands."⁹⁰

What differed between the United States and the Soviet Union, however, were the populations targeted by mail disruption. Both the authors of the 1989 report and an earlier Congressional hearing on the disruption of mail service in 1984 argued that Soviet Jewish people were disproportionately impacted by mail disruption.⁹¹ The report argued that these obstacles were indicative of an effort to "isolate the Jewish community in general, and the 'activists' in particular, from any contact with the outside world."⁹² The Soviet Union organized anti-Zionist campaigns throughout the latter half of the twentieth century, fueled in part by its negative relations with Israel during the Cold War. Anti-Zionist propaganda promoted historically anti-Semitic representations of Jewish people, which in turn provoked anti-Semitic attitudes in the Soviet Union, and these had negative effects on Jewish Soviets (Figure 4.3).⁹³ In a state where denunciations against a person's political or social loyalty could have deadly consequences, questioning Jewish loyalty through anti-Zionist campaigns was no small accusation. It is not unreasonable to assume that anti-Semitism could have been a reason for Shklovsky's difficulties with international correspondence, since he was at once a Jewish man and someone who might have been perceived as an activist.

After receiving Shklovsky's letter of frustration regarding the absence of updates on his book, Sagan responded, expressing distress at the news that Shklovsky had not received any of the four or five letters he had sent since May 1963. He acknowledged their communication challenges and showed empathy with Shklovsky's frustration.⁹⁴ Sagan let Shklovsky know that his book had been successfully translated, but further challenges had been presented regarding the figures and illustrations. The Soviet Union had not joined the Universal Copyright Convention, and therefore did not need to receive permissions from creators outside the Soviet Union to reproduce them in print. On the other hand, as Sagan pointed out in his letter to Shklovsky, the United States was a member of that convention; so tracking down and gaining permissions to reproduce

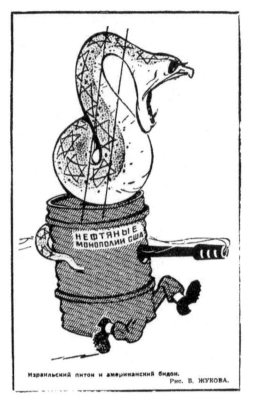

Figure 4.3 "The Israeli Python and the American Barrel." Cartoon by Boris Zhukov, *Pravda Vostoka*, February 11, 1968. Credit: Nir Yeshayahu, *The Israeli–Arab Conflict in Soviet Caricatures, 1967–1973*. Tel Aviv: Tcherikover Publishers, 1976.

the figures in the United States was a hefty task. Although this caused delay, it was not the main reason why the book was not yet ready for publication. Sagan was being mildly deceptive.

In an earlier letter, under the mistaken impression that Sagan had trained as a biologist and not as an astrophysicist, Shklovsky invited him to make edits and supplements to the parts of the book that covered biological subjects. He believed that Sagan's contributions in this area would be valuable, especially because, as a result of Lysenkoism, Shklovsky had limited access to a range of biology texts.[95] Decades later, Shklovsky claimed that Sagan "interpreted my request broadly," which, given what happened, might be considered an understatement.[96] In his letter explaining the delay, Sagan informed Shklovsky that he was compelled to introduce supplementary material to enhance the book's accessibility to western readers.[97] In his attempts to adapt the book to a United States readership, however, the young Carl Sagan admitted that he "found [himself] unable to resist the temptation to annotate the text, clarify concepts for the scientific layman, comment at length, and introduce new material" – until the English translation had about doubled in size.[98] This came as a shock to Shklovsky who, upon receiving the finalized copy of the English translation in the post, noticed that "on the cover were crammed the names of two authors: Shklovsky and Sagan."[99]

Shklovsky may have been understandably annoyed by this, but said of Sagan: "[he] showed a certain integrity; he left my text unchanged and set off his with little triangles" (Figure 4.4).[100] By "little triangles," Shklovsky was referring to the fact that Sagan, in his attempt to be clear about what constituted Shklovsky's original wording and what his own contributions were, enclosed his additional text within inverted delta signs. As a result of Sagan's attempts to delineate his own words and thoughts from Shklovsky's, the 1966 English edition, titled *Intelligent Life in the Universe*, reads in a somewhat awkward manner. It was neither a conversation between two scientists nor a streamlined narrative, but instead, as described by Sagan, "a peculiar kind of cooperative endeavor."[101] The disjointed organization highlighted the different perspectives of the two scientists, which served an extremely valuable purpose. On the one hand, it gave readers an insight into the different ways in which a Soviet and an American astrophysicist approached the problem of extraterrestrial communication. But, additionally and perhaps most crucially, it prevented Shklovsky from facing problems in the Soviet Union for what would have been viewed as undesirable arguments deployed within the book. Shklovsky was able to get the first edition of *Universe, Life, Mind* past the censors because of the fortuitous circumstances outlined earlier, but the new edition, with all of Sagan's cultural additions, would face greater problems. Sagan pointed out in the introduction to the book:

> As the reader might expect for a book written by two authors, one in the Soviet Union and one in the United States, there are occasional ideological differences. I have not tried to avoid these problems, but I also have not tried,

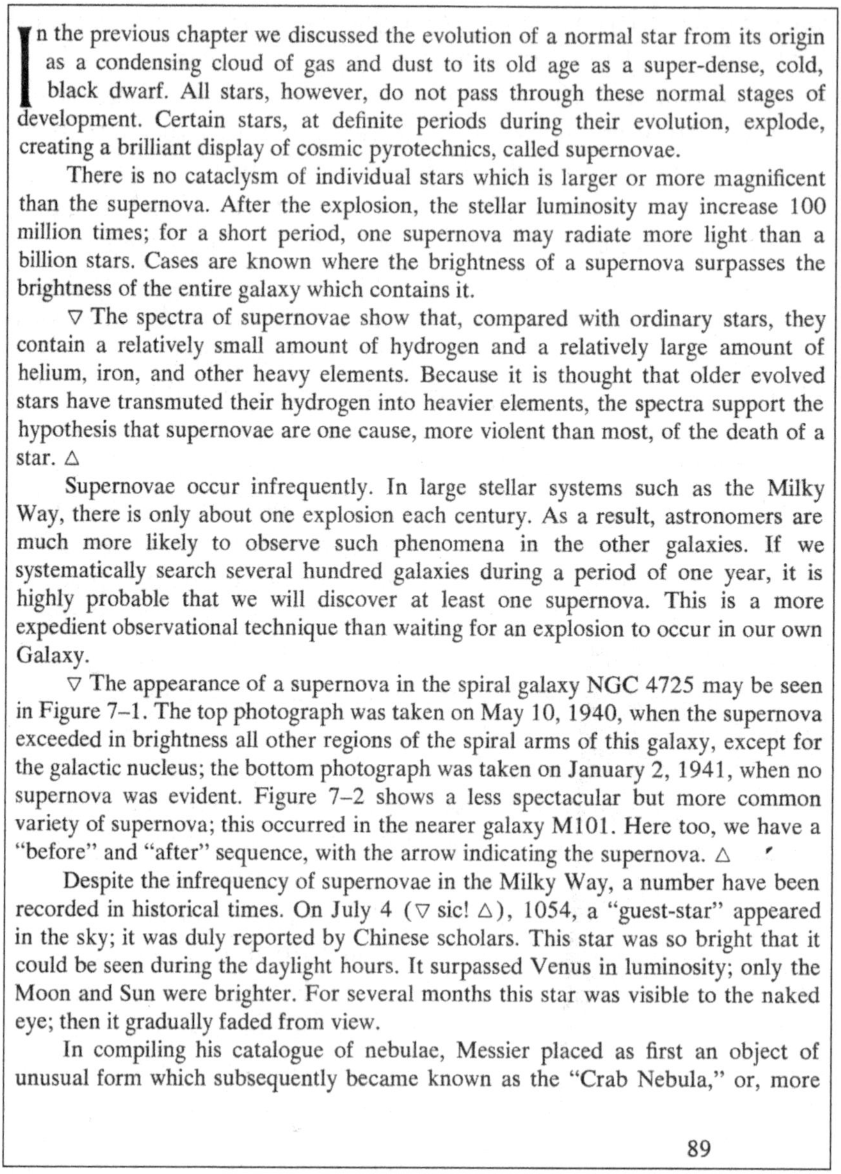

In the previous chapter we discussed the evolution of a normal star from its origin as a condensing cloud of gas and dust to its old age as a super-dense, cold, black dwarf. All stars, however, do not pass through these normal stages of development. Certain stars, at definite periods during their evolution, explode, creating a brilliant display of cosmic pyrotechnics, called supernovae.

There is no cataclysm of individual stars which is larger or more magnificent than the supernova. After the explosion, the stellar luminosity may increase 100 million times; for a short period, one supernova may radiate more light than a billion stars. Cases are known where the brightness of a supernova surpasses the brightness of the entire galaxy which contains it.

▽ The spectra of supernovae show that, compared with ordinary stars, they contain a relatively small amount of hydrogen and a relatively large amount of helium, iron, and other heavy elements. Because it is thought that older evolved stars have transmuted their hydrogen into heavier elements, the spectra support the hypothesis that supernovae are one cause, more violent than most, of the death of a star. △

Supernovae occur infrequently. In large stellar systems such as the Milky Way, there is only about one explosion each century. As a result, astronomers are much more likely to observe such phenomena in the other galaxies. If we systematically search several hundred galaxies during a period of one year, it is highly probable that we will discover at least one supernova. This is a more expedient observational technique than waiting for an explosion to occur in our own Galaxy.

▽ The appearance of a supernova in the spiral galaxy NGC 4725 may be seen in Figure 7–1. The top photograph was taken on May 10, 1940, when the supernova exceeded in brightness all other regions of the spiral arms of this galaxy, except for the galactic nucleus; the bottom photograph was taken on January 2, 1941, when no supernova was evident. Figure 7–2 shows a less spectacular but more common variety of supernova; this occurred in the nearer galaxy M101. Here too, we have a "before" and "after" sequence, with the arrow indicating the supernova. △

Despite the infrequency of supernovae in the Milky Way, a number have been recorded in historical times. On July 4 (▽ sic! △), 1054, a "guest-star" appeared in the sky; it was duly reported by Chinese scholars. This star was so bright that it could be seen during the daylight hours. It surpassed Venus in luminosity; only the Moon and Sun were brighter. For several months this star was visible to the naked eye; then it gradually faded from view.

In compiling his catalogue of nebulae, Messier placed as first an object of unusual form which subsequently became known as the "Crab Nebula," or, more

89

Figure 4.4 Excerpt from the first edition of *Intelligent Life in the Universe* (1966). Sagan made contributions to Shklovsky's original text by interjecting between " " symbols. In the second paragraph on this page, Sagan inserts a tongue-in-cheek *sic erat scriptum* ("so in the original"), to poke fun at the unintentional reference to the American Independence Day. Credit: the author.

in what is primarily a scientific work, to rebut each ideological assertion. When Shklovsky expresses his belief that lasting world peace is impossible while capitalism survives, or implies that lasers are being developed in the United States for their possible military applications alone, I have let the content of these statements stand, despite their political intent.[102]

Shklovsky later remarked on the benefit of Sagan's approach of separating their words with precision; he realized that "that my American 'co-author' had done me a priceless boon in distinguishing his text with triangles. Otherwise, our vigilant official 'readers' could have made things tough for me" (here Shklovsky was referring to the Soviet censors).[103] Clearly, the challenges faced by Shklovsky and Sagan in their attempts to work with each other across the Iron Curtain demonstrate that the CETI community found communicating on Earth posed nearly as big a challenge as communicating with extraterrestrials.

Intelligent Life in the Universe was one of the first CETI books to be published for a general readership, and quite possibly the first of its kind. It was written in the style of a popular science textbook aimed at teaching cosmology, planetary science, and principles of astrobiology to both university students and the general public. As for its structure, the book was divided into three parts: "The Universe," "Life in the Universe," and "Intelligent Life in the Universe." The first part exposed the reader to fundamental physics and chemistry principles that would set a foundation for the rest of the book. The second part introduced the reader to the science of cosmology and astrobiology, questioned how life developed on Earth, and explored the potential of the recently developed field of planetary science. The book's inclusion of planetary science was almost as unique as its content on extraterrestrial intelligence. In an interview published a few decades after the book's release, Sagan noted that "planetary science was considered disreputable" in the mid twentieth century.[104] "There was not a single other person working full-time on [planetary astronomy]," Sagan stated, except himself and his doctoral adviser at the University of Chicago, G. P. Kuiper. Indeed, according to Sagan, both the ongoing quest for extraterrestrial life and the goals of planetary astronomy, which were to investigate the topography and the chemical composition of other planets, were "considered nonsense" at the time of the book's publication.[105]

Sagan was not exaggerating when he said that many scientists in the mid twentieth century were resistant to new scientific fields that they thought were far-fetched or based on science fiction. When British Astronomer Royal Richard van der Riet Woolley was interviewed by the magazine *Time* in 1956, he said, in response to a query about the future of space exploration, "it's utter bilge."[106] Given that global investment and interest in space exploration would explode with the launch of Sputnik the very next year, such a statement reveals that there was little hope for more speculative disciplines like CETI and planetary astronomy. Therefore the choice to devote several chapters in *Intelligent Life in the Universe* to educating the public on these two issues was crucial to establishing them as serious domains of research.

The book's third section, which shared its name with the title of the book, was the most unusual one. The graphs and calculations that littered the previous sections all but vanish, replaced by cryptic photos of Assyrian cylinder relief sculptures and philosophical reflections on the possible theological consequences of contact with extraterrestrial intelligence. During its first year of publication, *Intelligent Life in the Universe* had approximately twenty thousand copies in circulation – a strong number for that time.[107] The reception of the book was mixed; it was popular with the public but met tepidly in the scientific community. One astronomer argued in a review that the book took "only [a] shadow of a factual approach," which he felt "served publicity" but was "misleading" and "unbelievable."[108] School teachers, on the other hand, praised the book, calling it "the most outstanding illustration of our contemporary thinking on this highly exciting subject," since it provided such extensive yet understandable scientific explanations of celestial phenomena; the book was assigned as a textbook for university courses.[109]

Despite its popularity upon publication, however, the book's success was short-lived. Today this is one of the few books written by Sagan that are no longer in print; and it has not been republished since 1998. This is not entirely surprising, given that today there are dozens of popular science books on the subject of extraterrestrial life (including this one). What sets *Intelligent Life in the Universe* apart from contemporary popular literature is the way in which it stands as a remarkable example of Cold War collaboration between Soviet and American scientists, and also as a demonstration of how CETI internationalism influenced the conduct of the scientists who practiced it. While the overt subject of the book is extraterrestrial life, the underlying thread that runs through it is humanity's place in the universe. While the book was firmly grounded in empiricism, it was also underpinned by a spiritual interest in the cosmic potential in humanity's future, in line with both the Soviet cosmist and the US frontier myths.

Furthermore, *Intelligent Life in the Universe* also addressed the longevity of extraterrestrial civilizations. Here the Cold War's connection with CETI rose to the surface; in the 1960s and 1970s, the United States alone possessed 1,054 nuclear missiles,[110] over ten times more than necessary to make the Earth hostile to human life.[111] For the first time in human civilization, human beings had the capacity to destroy their entire species in a nuclear war. This realization led Sagan and Shklovsky to make the following reflection:

> Another question of some relevance to our own time, and one whose interest is not restricted to the scientists alone, is this: Do technical civilizations tend to destroy themselves shortly after they become capable of interstellar radio communications?[112]

In the context of the Cold War, the question posed by both Soviet and American scientists exemplifies how CETI transcended physics and astronomy, prompting philosophical, historical, and sociological inquiries.

We have now caught a glimpse of some of the many challenges that faced astronomers during the Cold War, especially those in the Soviet Union. But we have also seen how internationalism in CETI fueled cooperation across the Iron Curtain. This internationalism was the product of a great optimism upheld by CETI proponents. In an interview near the end of his life, Shklovsky was asked whether he had experienced any difficulties or tragedies in his life. Surprisingly, he responded: "Tragedy? No, absolute [*sic*] 100 percent satisfied. Life is a blue dream."[113] This response was remarkable, as we have seen that Shklovsky's life was riddled with hardships – he experienced anti-Semitism, travel bans, mail interference, and even threats of death. He spent much of his life in poverty. Before the interview concluded, Shklovsky was asked: "Is there anything you would like to have people in the future know about your life?" He responded by contradicting his earlier statement: "I am *not* satisfied completely with my life." The interview continued:

> INTERVIEWER: What would you like people in the future to know about your feelings?
> SHKLOVSKY: [pause] A little more degree of freedom.[114]

Shklovsky's answers highlight the paradox of internationalism in CETI: it exposed the systemic challenges and tensions of the Cold War era, while simultaneously fueling a strong determination to overcome those barriers and fostering hope in a utopian future. In other words, CETI scientists worked in the hope of achieving a little more freedom.

5

Why Can't We Be Friends?
Messaging Extraterrestrial and Terrestrial Intelligence

It is a dream pulled straight from the pages of a science fiction novel. A radio astronomer reviews the data from her observations when, suddenly, she notices something strange. How strange is it? Maybe it's a narrowband signal, a signal that is much more compressed in frequency than we might expect to see from natural sources, and it clearly isn't local. Or perhaps she sees a signal that is significantly stronger or more persistent than she would expect to find in nature. Upon further investigation, the astronomer and her team discover a message hidden in the signal. It might be simple, pulsed to indicate a string of primary numbers, like the signal from the film *Contact*. Or maybe, as Carl Sagan once put it, there is an *Encyclopedia Galactica* encoded in the signal, ready to transform human civilization via a wealth of information and knowledge about the universe. But one way or another, there is no doubt: the signal is as clear as day. Human beings have finally detected a message from extraterrestrial intelligence.

In contemporary SETI circles, this fantasy is often described as the anticipated champagne moment – the definitive moment to pop the bottles we have been saving up for the day we finally detect evidence of extraterrestrial intelligence. While there is an expansive literature speculating on how the champagne moment might be arrived at, at least in twentieth-century CETI there was a general expectation that any signal we detect would be a directed signal (in other words, a deliberate message from extraterrestrial intelligence beamed in our direction). Partly this expectation stemmed from the limitations of receiver sensitivity in the twentieth century – it would be too difficult to detect faint, unintentional signals with the tools of the time – but also from an underlying assumption that a communicative technological species such as our own would be likely to send a targeted message. But just how often have we been sending signals of our own?

The era of CETI – which, to repeat here, stands for communication with extraterrestrial intelligence (as opposed to the more contemporary SETI, search for extraterrestrial intelligence) – should be considered a historically unique era in the larger history of human speculation on life in the cosmos. Clearly there are strong connections between the Cold War and the development of CETI, and communication underlined every aspect of these connections.

And yet, the term CETI was somewhat of a misnomer. As I have shown thus far in the book, the process of CETI was largely affected by communication on Earth. But little attention was given to *making* attempts at communication with extraterrestrial intelligence – actual CETI. This is no mistake; as CETI scientists later realized, successful communication with extraterrestrial intelligence was fraught with major challenges. Few attempts at direct communication with extraterrestrial intelligence were made during the Cold War period. In fact, there are only five messages that can be considered candidates for the earliest attempts at sending communications to extraterrestrial intelligence. Evidently, the C in CETI might stand more for communication between the United States and the Soviet Union than it does with extraterrestrials.

The Morse Message

Shortly before the turn of the year in 1962, Venus became the first planet in our solar system to receive a deliberate radio message from the planet Earth. The message, now known simply as "the Morse message," was sent using the Soviet planetary radar complex in Yevpatoria shortly after its construction in the early 1960s. The Soviet Union became the first country to use radar facilities to send a message into space. The Morse message is considered, by and large, the first attempt to communicate with extraterrestrial intelligence.[1] The science fiction magazine *Lightspeed* describes the Morse message as "one of the first radio signals intended specifically for interstellar intelligence."[2] Matt Reynolds at *Wired* describes it as the moment when "humankind sent its first message to extraterrestrials."[3] Dumas, the astrophysicist at SETI League just mentioned (n. 1), created a list in which the Morse message is credited with being the first "real messaging extraterrestrial intelligence [broadcast]" and the "first message to extraterrestrials."[4] Despite the public and scientific consensus, how-

Figure 5.1 This image shows a newspaper clipping describing the success of the Soviet Morse message. Credit: "Words to the Cosmos: Peace, Lenin, USSR." *Krasnaia Zvezda (Red Star)*, December 30, 1962.

ever, it is important not to forget that the Morse message was a product of the Cold War and in reality was not a message to extraterrestrials at all.

The message was designed in Morse code using frequency manipulations and was sent in two stages. The first part of the message, which simply contained the word мир ("world," "peace"), was sent to Venus and reflected back to Earth, where it was received at Yevpatoria. The second part contained the words Ленин ("Lenin") and СССР ("USSR") and was similarly transmitted to and received from Venus using the same method (Figure 5.1). The Soviet newspaper *Krasnaia Zvezda* (*Red Star*) excitedly reported news of the message in December 1962:

> The Institute of Radio Engineering and Electrical Engineering of the Academy of Sciences of the USSR informed [us] of a new victory for Soviet science and technology ... For the first time in the history of mankind, Soviet scientists succeeded in carrying out radio communications through to the planet Venus.[5]

The Soviet Morse Message was the first radio message deliberately sent into space using a radio astronomy facility; but the intended recipient of the message was likely Earthlings, not Venusians. While this is an assumption, it is a safe one: the three scientists involved in creating and sending the message, Drs. Vladimir F. Morozov, Oleg N. Rghiga, and Vladimir M. Dubrovin, were not active members of the growing Soviet CETI community. The Soviet Morse message seems uniquely designed for a human audience: it was sent in Russian Morse code, a system of communication that approximates traditional Morse code in that it encodes the Cyrillic alphabet in a standardized sequence of two different signal durations. An alien civilization might be able to decode a base-2 communication method that centered on numbers, but probably not one that required a preexisting understanding of the Cyrillic alphabet and of the Russian language.

Additionally, the message was nationalistic in nature. Rather than attempting to send a greeting or a universal message that represented Earth or humankind, the Soviet Morse message highlighted its own country (the Soviet Union) and its foundational belief system and main ideological figures (Lenin). Why did the Soviet Union send such a message? As already noted, in the early 1960s the space sciences were highly mythologized as national projects, both in the United States and in the Soviet Union. In the United States, space was mythologized as an extension of the belief in manifest destiny, which professed that the United States had a destiny to expand "from sea to shining sea" and that this exercise in exploration, conquest, and expansion could extend into the final frontier – outer space. The astronaut became a sort of cowboy and the aesthetics of the American West blended into the space age. In the 1960s, children ran around with toy laser guns that had replaced their toy pistols and pretended to shoot their friends, now imaginary space aliens instead of American Indians. The "space western" became a subgenre of science fiction found in the 1960s in comic books, in pulp magazines, in TV shows, and even on the silver screen. The US space program validated the ideological principles of rugged individualism: astronauts had "the right stuff."

The US space age complemented capitalism – NASA partnered with private companies in its moonshot. In 1968, the year before Apollo 11 landed on the Moon, NASA spent $453,000,000 on private sector service contracts to its facilities and worked with approximately 28,500 private contractors in that year.⁶ Walt Disney built the space age Tomorrowland, which promised a future built through innovation and investment in technology and business. The initial opening of Tomorrowland was somewhat of a corporate showcase, with companies such as Monsanto, Richfield Oil, and American Motors hosting showcases in the theme park.⁷ The United States characterized its choice to go to the Moon as fitting in with this individualistic, hardworking frontier ethic. Kennedy asked:

> Why choose this as our goal? And they may well ask why climb the highest mountain? Why, 35 years ago, fly the Atlantic? Why does Rice play Texas? We choose to go to the Moon in this decade and do the other things, not because they are easy, but because they are hard.⁸

American football games or moonshots . . . space was simply a new part of a national tradition (Figure 5.2).

In the Soviet Union, outer space was characterized by a similar, yet different mythologizing. The cosmonaut underwent a sort of apotheosis, becoming a larger than life figure that represented not simply an individual person but the ideal of the "new Soviet man." Cosmonauts were described at the time as people who demonstrated "in action all the invaluable qualities of the Soviet character, which Lenin's party has been cultivating for decades."⁹ They were picked in part for their personal lives: Yuri Gagarin, the first human being to enter outer space and orbit the Earth, exemplified the Soviet proletariat, having come from a working-class background (his father a carpenter, his mother a dairy farmer). The Soviet Union created visual propaganda that envisioned a utopian communist future in space (Figure 5.3). Space was not the "final frontier" in need of individual or business enterprise, but rather the next step in the communist revolution, which would lead humanity to a peaceful and unified cosmic utopia. Soviet space posters claimed "Socialism is our launching pad!" and portrayed a destiny among the stars, led by the Soviet Union.

These national myths served many purposes: to galvanize citizens into supporting space efforts, to garner public funding (in the case of the United States), and to justify these pursuits as being part of a long

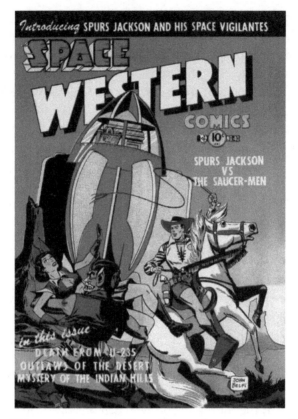

Figure 5.2 This comic book titled *Space Western* features a dynamic scene of a futuristic cowboys chasing an alien villain. The blend of traditional western motifs with futuristic space exploration was a common pop culture theme during the Cold War era. Credit: Archive PL / Alamy Stock Photo.

tradition aligned with their respective country's history. But there was a further reason for creating grand narratives and myths around space efforts. The term "Cold War" now generally refers to a specific conflict, namely the one between the United States and the Soviet Union in the mid twentieth century. But literally it refers to the fact that the conflict was largely indirect (as opposed to the "hot" wars fought by direct combat). The United States and the Soviet Union characterized their battle as being as an ideological and existential one, which required weapons beyond those used in traditional warfare, but also as extending into the realms of cultural and psychological warfare.[10] This psychologi-

Figure 5.3 This Soviet poster from 1961 states: "Socialism is our launching pad!" It highlights the fusion of Soviet economic and political goals with a future of space exploration.
Credit: The Protected Art Archive / Alamy Stock Photo.

cal warfare played out in various propaganda campaigns. In the United States this meant stemming the tide of communism; in the Soviet Union, it meant ensuring the spread of communism and undermining faith in democratic capitalism. Many of these campaigns targeted what was then called "the third world."

Now largely seen as a derogatory term for economically developing or previously colonized countries, the term "third world" had a specific meaning during the Cold War. In the decade or so after the end of World War II, 37 previously colonized nations gained independence. The United States and the Soviet Union engaged in fierce competition for control and influence over these newly independent nations. This battle was largely ideological – a grand struggle between capitalist democracy and communism for the soul of the world. Soviet cosmonauts went on political world tours, as part of this effort to persuade and influence the third world. Yuri Gagarin made approximately 200 trips to foreign countries between 1961 and 1968.[11] In the United States, a radio program called Voice of America delivered American culture and news to audiences in over 100 countries, broadcasting in 46 different languages.[12]

Considering this historical context, we can now see why the Soviet Morse message was promotional and nationalistic in nature. This much is clear from the *Krasnaia Zvezda* report. Although the newspaper opened with exciting news about a message to Venus and displayed the strip chart recording of the message prominently in print, most of the article was dedicated to the new radio astronomy and radar facilities at Yevpatoria, which had already made the Soviet Union the first nation to begin radar observations of the planet Mercury. We can conclude that the message "PEACE, LENIN, USSR" – the first such message to be sent to another world – was not a message for extraterrestrials. It was a message for Earth. But, although it was not a genuine attempt at CETI, the use of the Yevpatoria array in this manner did suggest the real possibility of engaging in cosmic conversation for the first time in human history.

Pennants, Plaques, and Puritans

The dispatch of the Morse message was not the Soviet Union's first attempt to announce its existence to a celestial body. A few years earlier, in 1959, the Soviet Union launched Luna-2, the first spacecraft to reach the moon. The spacecraft carried two titanium spheres that contained an explosive charge designed to burst on impact and to scatter numerous pentagon-shaped pennants across the surface of the Moon (Figure 5.4). The pennants were simple in design; they showed the Soviet state seal (a wreath of grain encompassing a hammer and sickle) and the Russian acronym CCCP ("USSR") accompanied by the date of the mission, September 1959. It is not known whether the pennants reached the lunar surface successfully: Luna-2 achieved what is known as a "hard landing" (basically, it crashed).[13] Its impact was believed to have a relative velocity

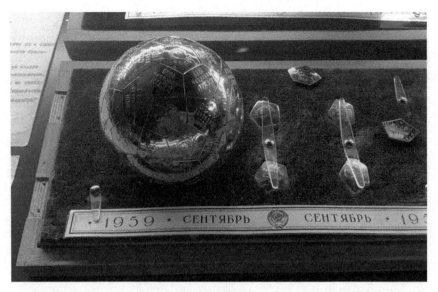

Figure 5.4 These pennants, replicas from the Luna mission, are held at the Russian Museum of Cosmonautics in Moscow. Credit: Rebecca Charbonneau.

of 3.3 kilometers per second, so many believe that the spacecraft and the pennants were vaporized.[14] Just as with the Morse message, the pennants were not designed to communicate with extraterrestrial intelligence but rather to signal technological and political power to those who watched from Earth (Bernard Lovell tracked the transmission from the probe and played a recording to global press).[15] But the Soviet pennants became the first in a long-lasting tradition of sending symbols and other forms of terrestrial communications into space. During the Space Race, these symbols were decidedly political (e.g. the American flag planted on the lunar surface) or commemorative (e.g. the Fallen Astronaut, an aluminum statue placed on the Moon by the crew of Apollo 15 to honor those who died in the Space Race).

But in the post-Space Race era, the intended audience for messages in space seemingly turned from terrestrials to extraterrestrials. In the early 1970s, Carl Sagan was working at NASA on the Pioneer program, which was set to launch two solar system probes, Pioneer 10 and Pioneer 11. Both were part of the larger NASA Pioneer mission, which was a series of eight spacecraft missions aimed at exploring our solar system. Pioneer 10 launched on March 2, 1972 and became the first spacecraft to make direct observations of Jupiter; Pioneer 11 launched on April 6, 1973 and became the first human-made object to fly past Saturn. In addition to these scientific firsts, Pioneer 10 and Pioneer 11 are distinct from the rest of the Pioneer missions by virtue of being the first two spacecraft to carry a message from our planet to extraterrestrial intelligence.

Living up to their name, Pioneer 10 and 11 would not only explore new worlds but attempt to serve as ambassadors to any intelligent life they might encounter. A gold-plated aluminum plaque was placed on each probe with a message from Earth, designed by Sagan and Drake (Figure 5.5).[16] The plaque design included four major symbols.

Underneath the representation of the hydrogen atom was a map designed by Frank Drake, intended to give an approximate location of

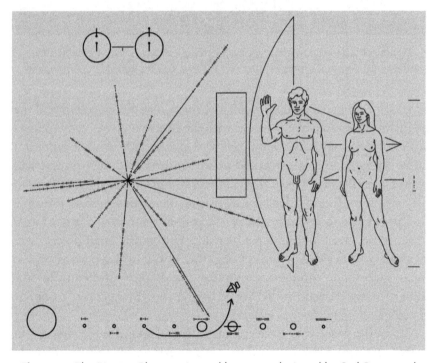

Figure 5.5 The Pioneer Plaque, pictured here, was designed by Carl Sagan and Frank Drake with artwork by Linda Salzman-Sagan. It was placed on board the 1972 Pioneer 10 and 1973 Pioneer 11 spacecrafts. Credit: NASA / Wikimedia Commons. At the top there was a schematic representation of the hyperfine transition of hydrogen (referring to the significance of the previously discussed hydrogen line for extraterrestrial communication).

the Sun. This was accomplished by creating a radial pattern of 15 lines that emanated from the same source (the Sun). Of these, 14 were interrupted by a series of dashes – a binary representation of the periods of known pulsars (highly magnetized neutron stars, which can be seen to "blink" like a lighthouse when observed in the radio from Earth). Each line was a different length, which was meant to show the relative distance of the pulsars from the sun. An extraterrestrial civilization that knew of those pulsars might be able to roughly approximate which stellar system the Pioneer craft originated from. The plaque also displayed a symbolic representation of our solar system, showing what was then known as the nine major planets and the Sun. A symbol representing the Pioneer probe had an arrow indicating that the probe was launched from the third planet from the Sun – Earth.

But the most notable as well as controversial feature on the plaque was the prominently displayed illustration of two human beings. The illustration was a line drawing of a man and woman in the nude, standing side by side. The male figure is depicted waving "hello" in greeting, while the woman stands beside him, arms to her side, in contrapposto. The illustration was designed by Carl and his then wife, Linda Salzman-Sagan, who made the drawings. The Sagans wanted to generate an inclusive portrait of humanity, representative of all its members – a cosmic mirror image. In later autobiographical writings, Sagan claimed that they tried to make the man and the woman look "panracial" by using ambiguous features such as wavy hair, no skin coloring, and simply rendered facial features. The figures were also inspired by classical Greek imagery, particularly Greek statuary.

Despite these efforts to create a universal depiction of humanity, however, the images of the humans on the Pioneer plaque caused outrage. Feminist groups were especially angry: while the depiction of the man was anatomically correct, the woman's genitalia were conspicuously missing. Sagan later gave a justification:

> the principal feminine criticism is that the woman is drawn incomplete – that is, without any hint of external genitalia. The decision to omit a very short line in this diagram was made partly because conventional representation in Greek statuary omits it. But there was another reason: Our desire to see the message successfully launched on Pioneer 10.[17]

Sagan meant that there was concern that the depiction of a culturally taboo feature like the female genitalia might prevent the plaque from being approved by what he called NASA's "scientific–political hierarchy."[18] Robert Kraemer, then director of NASA's planetary programs, further explained in his memoir:

> When the plaque design was submitted to NASA headquarters for approval, I must confess that I was a bit nervous about it. Linda was a skilled artist and her naked human figures were very detailed and realistic, as they needed to be. It seems a bit silly today, but at the time I feared some taxpayers, the true owners of the spacecraft, might label it pornographic. My boss, John Naugle, had no such fears and approved the design but with one compromise of erasing the short line indicating the woman's vulva. (The poor extraterrestrials are probably going to be puzzled by the functional differences in anatomy between the two figures.)[19]

Although feminist groups were infuriated by NASA's concessions to puritan taboos related to the female body, Kraemer was correct in his prediction that others would be repulsed by what they perceived as NASA sending "smut" to space. In a letter to the editor of the *Los Angeles Times*, one man drew a comparison between the Pioneer plaque and the "bombardment of pornography through the media of film and smut magazines," lamenting that even our "space agency officials have found it necessary to spread this filth even beyond our own solar system."[20] The nudity on the plaque was mocked in newspapers and cartoons around the country; one cartoonist depicted a resident of Jupiter remarking: "the Earth people are evidently very similar to us here on Jupiter . . . except that they don't wear any clothes!" The nudity on the Pioneer plaque laid bare many points of cultural tension within the United States, ultimately highlighting our differences instead of achieving the universal depiction of humanity it sought to produce.

* * *

Interestingly, after the Morse message, the second ever deliberate attempt to send a message to other worlds using radio astronomy infrastructure was actually an attempt at correcting the Pioneer plaque. In 1982, frustrated that he had been rejected for an arts fellowship, artist Joe Davis walked into the MIT Center for Advanced Visual Studies uninvited (which must have been quite a sight: Joe Davis had lost one of his legs only a few years earlier and favored a pirate-style peg leg for a prosthesis). The secretary called the cops. About an hour later, Davis left the Center, having been awarded an academic fellowship (believe me, he's an enigmatic and persuasive fellow). One of Davis' first projects at MIT was a semi-covert operation designed to rectify what he believed to be an offensive image on the Pioneer plaque. Davis later stated: "We sent pictures of a man and a Barbie doll to space."[21]

The project, a conceptual artwork titled Poetica Vaginal (1986), aimed to correct the Pioneer plaque's missing element. It was a highly collaborative project. Davis worked with MIT engineers to build a "vaginal detector," which "consisted of a water-filled polyallomer centrifuge tube mounted on a hard nylon base that contained a very sensitive pressure transducer."[22] Dancers from the Boston Ballet and several other female volunteers then "invaginated" (Davis' preferred term) the detector, which was able to record their voices, heartbeats, and vaginal contractions.[23]

Davis and his team then used electronic music software to generate harmonics of the contractions that matched the frequencies of human speech. A collaborating linguist then drew a corresponding English phonetic map of these sounds, creating a "message" of sorts from the vaginal movement, as if the vaginas could communicate for themselves. And, finally, a digital map of the detector output was put together. Thus, as Davis later described, "three forms of the message were simultaneously generated: (1) an analog signal directly generated by vaginal contractions; (2) a digital map of same and (3) voice (English phoenetic maps of vaginal contractions)." Electrical engineers, mechanical engineers, and artists on the team then worked to build the appropriate "Vaginal Excursion Module," assembled at MIT and designed to transmit vaginal signals recorded on an audio tape.[24] Using the Millstone radar at MIT's Haystack Observatory, these recordings were transmitted in the direction of Tau Ceti and Epsilon Eridani – the same two-star systems Drake observed in Project Ozma in 1960.

The operation was necessarily secretive: the Millstone Hill Steerable Antenna has long played a key role in the US deep space surveillance program, largely because of its usefulness in monitoring the near-space environment, particularly in tracking satellites. Because of this use, the Millstone radar was partly operated by the US Air Force. But the project, despite best efforts, did not remain covert. In our interview, Davis noted, "we didn't say 'we're transmitting vaginal contractions.' [We said] we're making an artistic transmission . . . this is a project in art and science. But somebody spilled the beans." As a result, approximately twenty minutes into the broadcast, the Air Force found out about the contents of Davis' artwork. Davis says he received a heated phone call from the Millstone project group leader, a US Air Force Colonel, during Poetica Vaginal's transmission:

> some colonel from the Air Force called me up and said, "You have to shut down now!" and I went, "why?" And he said, "You know why!" And I said, "No, I don't, tell me why." . . . He never said.

Davis uses this episode to highlight how messages we send to space are burdened by terrestrial baggage. He argues that this is an intrinsic element of CETI: "By making this attempt to communicate with the other . . . we're really communicating with ourselves."[25] By this point in the book, such a simple message should strike the reader as mundane and obvious. But my goal is not only to persuade you of the "cosmic mirror" elements of CETI (which Davis correctly points out) but also to show how all this was a product of the Cold War. There are two ways in which we can see this in the case of the reaction to the Pioneer plaque and Poetica Vaginal.

First and most obviously, there is the duality of the scientific–military infrastructure, sometimes known as the military–industrial complex. The Millstone radar facility, in addition to being used by MIT for scientific purposes, was also used by the US military for detecting and monitoring objects located in deep space and on geosynchronous orbits. The Millstone radar was one of the first radar facilities to detect the Sputnik I satellite. But, second and more covertly, the cultural environment of the mid twentieth century highlights CETI's Cold War connection. In 1971, the year before the Pioneer plaque was first launched aboard its spacecraft, the US Supreme Court affirmed, in *United States v. Reidel*, that laws forbidding the distribution of obscene materials were constitutional. That ruling represented the culmination of a long-running struggle to define and handle "obscene" materials.

Just seven years earlier, in 1964, Justice Potter Stewart famously defined the test for pornography in his concurring opinion on *Jacobellis v. Ohio*, an obscenity case: "I know it when I see it."[26] The US Supreme Court first addressed the issue of obscenity in 1957, in *Roth v. United States*, which found that obscene materials were not "within the area of constitutionally protected speech or press."[27] It is not a coincidence that these cases began in the 1950s, in the early days of the Cold War. In the 1950s the United States was furiously redefining itself against the Soviet Union and, in doing so, forming a new national identity centered on Christian ideals, democracy, and capitalism. In 1956, the year before the first Supreme Court obscenity case, the United States changed the official motto on its Great Seal from *E pluribus unum* ("Out of many, one") to "In God We Trust," despite the nation's founding ideal of separation between church and state. In the same decade, fear of the spread of communism led to highly publicized probes, pioneered by US Senator Joseph McCarthy, into alleged communist infiltration of the US government.

But the investigations did not stop with alleged communists. In the 1950s and 1960s, queer individuals, or those who were perceived as gender non-conforming, were dismissed and barred from government service because of the perception that they were more likely to be communist sympathizers – a moral panic now known as "the Lavender Scare." Queer historian Whitney Strub has argued that obscenity laws and the Lavender Scare were intimately connected to the larger Cold War cultural environment. He has argued that obscenity laws "fit wholly into the political matrices of the Cold War . . . Sexuality took on political hues in the convoluted but tightly organized framework, as a specific regime" that was "organized around the ideal American Cold War family."[28] These laws did not only target members of the LGBTQ+ community but also aimed to create a rigid framework of social conformity, enforcing traditional gender roles and norms to reinforce the idealized American Cold War family. Individuals perceived as engaging in acts of "sex perversion" were targeted through obscenity laws, which encompassed pornography makers and distributors, stand-up comedians, and even ordinary women who wore swimsuits or clothing deemed "obscene." As he connects the Lavender Scare and obscenity laws to the larger cultural context of the Cold War, Strub argues that the regulation of sexuality served as part of a broader effort to "strengthen our home fronts" and "preserve the nation," illustrating the extent to which the fear of deviating from established gender and sexual norms shaped the era.[29] Rather than creating a cosmic mirror, Davis argues that the Pioneer plaque created "a picture of our own intolerance" through its reflection of the fear and bias surrounding sexuality during the Cold War period.[30]

Poetica Vaginal is generally not taken seriously by the SETI community. It is usually not listed among the numerous attempts at radio communication with extraterrestrial intelligence.[31] According to Davis, at the time the artwork was created, it was met with derision and disgust. When I speak of this project at SETI meetings, I often encounter laughter (it's ok, it is a bit funny). But in our recent interview Davis defended his project: "They think that it wasn't serious, but it was quite serious. And it was feminist."[32]

Davis' art is typically classed by the art world as being part of the BioArt movement – an art practice in which artists incorporate biological components (e.g. living tissue, bacterial cultures, genetics, or biological processes) into their art. But a novel way to examine his work is through the lens of the abject art movement.

Abject art arose as a response to the censorship faced by artists at the hands of the religious right in the early 1980s, around the time when Davis created Poetica Vaginal. Continuing the anti-obscenity laws that proliferated during the mid twentieth century, prominent political figures vociferously criticized artists such as Robert Mapplethorpe, primarily targeting their use of queer imagery, but also artists who depicted the female body in any manner that was not considered "classical" or pandering to the traditional male gaze. Abjection, a concept popularized by the renowned French philosopher and psychoanalyst Julia Kristeva, is defined by elements that disrupt "identity, system, [and] order."[33] The abject compels individuals to experience sensations of repulsion and unease, urging them to reflect on the reasons behind such emotions. In doing so, it subverts the conventional belief that art exists solely for the purpose of beauty.

Abject artists often incorporated elements designed to evoke that sense of disgust, sometimes using human bodily fluids as media, or otherwise depicting the human form in disquieting or "ugly" ways. Juliette Austin tells us that abject art "has a strong feminist context, in that female bodily functions in particular are 'abjected' by a patriarchal social order."[34] One famous example of feminist abject art are the works of Cindy Sherman, who in the 1970s and 1980s created self-portraits made up into disturbing caricatures, such as of a nude corpse or of a brutalized woman. Her abject photographs are horrific, repulsive, and a far cry from the idealized image of the female nude popularized throughout western art history. Her images compel viewers to step out of their expectations of how to view the female form and force them to confront the dark side of the human, and specifically the female, experience. They put into question the idealized woman and ask the viewer to consider a harsher reality.

The word "abject" comes from the latin verb *abicere*, "to throw away, cast off." Poetica Vaginal has many of the characteristics of abject art, given its use of human body parts, the repulsion and anger it evoked, and the feminist intent of its message. But in some sense it also invites us to consider the Pioneer plaque as a work of abject art: the plaque literally portrays a woman who is disfigured or has had a part of her body cast off. It provides an incomplete and corrupted view of our world, not a cosmic mirror image of humanity.

The Expectation of Man*kind*

Joe Davis' Poetica Vaginal could be considered by some as an example of messaging extraterrestrial intelligence but, as Davis himself has pointed out about his projects, "it's [really] about communicating with ourselves."[35] As we have seen so far, there have been few examples of using radio telescopes and radar facilities to send messages to extraterrestrials with a truly extraterrestrial audience in mind. The first messages sent into outer space had decidedly terrestrial audiences in mind.

Davis' message was unique in that it functioned as a critique rather than a show of power or attempt at universalizing humanity. Specifically, Poctica Vaginal highlighted the gender dynamics at play in the messages we send. It forced the CETI community and its audiences to reckon with sex and gender inequity.

This is not to say that the CETI community was intrinsically antifeminist. Many CETI pioneers were advocates for women, including Sagan and Drake, despite the issues with the Pioneer plaque. Frank Drake was an avid supporter of women in science. When he was planning Project Ozma in 1960, Drake was also tasked with hiring summer students for the Observatory. Since 1959, the National Radio Astronomy

Observatory (NRAO) has hired undergraduate students to work over the summer with scientists and to conduct research projects. In the first year of the program, the students were all male. But in 1960 Drake decided to include women in the summer student cohort. He recorded that only two of the 12 students he had admitted into the program were women, but this was enough to evoke outrage at the Observatory, particularly from John Findlay, a more senior scientific staff member.

Drake recalled that Findlay was "furious" at his decision to admit female students and later paraphrased him as stating: "What are you doing, Drake, to do such a dumb thing? This is a total waste of resources and contrary to all tradition!"[36] According to Drake, Findlay believed that investing in female students was a waste of resources; they would eventually get married, become mothers, and not contribute to astronomy.

But Drake knew from experience that women could be excellent astronomers. His thesis adviser at Harvard had been none other than Cecilia Payne-Gaposhkin. Payne-Gaposhkin was the first person to earn a PhD in astronomy from Radcliffe College – Harvard's women-only institution, founded in 1879 (at a time when women were not admitted into Harvard). Later on Payne-Gaposhkin became the first woman to be promoted to full professor from within the Harvard Faculty of Arts and Sciences. Her research on stellar composition revealed that the stars are primarily composed of hydrogen and helium, fundamentally shaping our understanding of astrophysics and the elemental composition of celestial bodies.

Drake thus persevered in his decision to admit female students, and in the summer of 1960 the two students, Ellen Gundermann and Margaret Hurley, came to Green Bank.[37] Drake did not just admit the young women, but supervised them himself. Ellen and Margaret were the only other people in the control room when Project Ozma observations were conducted, so they made history by assisting Drake in the first radio astronomy CETI experiment.

Carl Sagan was, similarly, a staunch supporter of women's rights. He authored essays in defense of reproductive rights, wrote an impassioned letter to his fellow members of the Explorer's Club – a professional society for the promotion of scientific exploration – asking for the admittance of women, and wrote a science fiction novel that featured a female scientist as the protagonist and a female president of the United States as a supporting character. It was he and Linda Salzman-Sagan, his then wife, who designed the figures on the Pioneer plaque, and the accusations of sexism took Linda by surprise. She considered herself a "liberated woman" and had not intended to present women as unequal to men on the plaque.[38]

This is a critical piece of information: CETI scientists were not necessarily sexist – in fact many of them considered themselves supporters of women's rights. They often went to great lengths to reflect on gender dynamics in their messages and careers at large. They included women's voices and images on and in the messages they constructed. But they were nevertheless still limited by the cultural milieu they operated in. For example, in reminiscing on his decision to admit women into the NRAO summer program, Drake wrote of one of the students, Ellen:

> [She] vindicated my judgment by going on to take her PhD in astronomy at Harvard. She later joined in the discovery of the first molecule to be found in interstellar space, and helped establish the existence of molecular line astronomy, which is now a major field of study. Eventually, however, she did what Findlay predicted would happen: She got married and went on the mommy track.[39]

This represents a limited appreciation of the merits of working to include women in science, and one not entirely fair to Ellen Gundermann.

Ellen did indeed go on to marry and have a child, but not before remarkable scientific accomplishments. As a graduate student at Harvard, she did more than just detect a molecule in space: seemingly she made the first detection of a cosmic maser. This is an astrophysical phenomenon similar to that of lasers, except that, instead of emitting visible light, masers emit microwave radiation. Masers would shortly become a major subject of astrophysical research. Ellen was apparently scooped before she could publish her results, which can be found in her Harvard thesis.[40] After her time at Harvard she went on to become a postdoctoral fellow at the California Institute of Technology, a prestigious accomplishment practically unheard of for women of that time.

She then married a fellow astronomer, Harry Hardebeck, who was hired as an astronomer at CalTech's Owens Valley Radio Astronomy Observatory. Although it is difficult to say why Ellen left astronomy shortly thereafter (I have not yet had an opportunity to interview her or her family), her story fits in with a common pattern for women in science; inevitably, their husband's career takes priority and they leave (or are forced to leave) the field. Nevertheless, Ellen did not simply "join the mommy track," as Findlay patronizingly predicted. She later found a new career working as the air pollution control officer at the Great Basin Unified Air Pollution Control District, a California regional government agency that seeks to protect the people and the environment of Alpine, Mono, and Inyo Counties from the harmful effects of air pollution. Furthermore, Ellen's daughter, Jean Hardebeck, would grow up to become a scientist herself. Jean received her PhD degree at Caltech and became an award-winning geophysicist. She studies Earth's fault lines, aiding our understanding of the efficacy of earthquake prediction models. The successes of these women in science would likely not have been possible without Drake's insistence to admit a young undergraduate girl into NRAO's summer program in 1960, even though he was ultimately disappointed by Ellen's decision to leave astronomy for motherhood.

CETI scientists like Frank Drake did important work to combat gender bias, even though at the same time they inevitably reproduced it. This is not a condemnation of individuals but rather a key consideration: it is often impossible to truly escape our cultural biases and, even when we attempt to embrace universality, we inevitably betray the markers of our limited perspectives.

Keeping this in mind, if we wish to continue our examination of the history of how members of the scientific community used radio astronomy infrastructure to make contact with extraterrestrial intelligence, one could argue that the first genuine attempt at radio contact was a message designed and sent by Frank Drake in 1974 (Figure 5.6). Drake was then the director of the National Astronomy and Ionosphere Center, which operated the Arecibo telescope. The Arecibo telescope was a large radio telescope located in Arecibo, Puerto Rico. It was one of the world's most powerful and iconic radio telescopes, featuring a 1,000-foot spherical reflector dish and powerful radar capabilities and suspended by cables above a natural sinkhole. In 1974 work on upgrades to the telescope had just been completed and Drake was tasked with commemorating what was essentially a new instrument. Drake credits his administrative assistant, Jane Allen, for coming up with the idea to send a message to

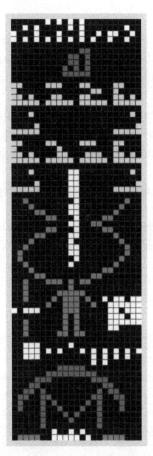

Figure 5.6 The Arecibo Message was sent on November 16, 1974 by the Arecibo radio telescope in Puerto Rico.
Credit: Arne Nordmann / Wikimedia Commons.

extraterrestrial intelligence during the telescope's dedication ceremony – another example of his taking steps to acknowledge and uplift the contributions of women.[41] Galvanized by her idea, Drake set to work on designing a radio message.

Drake's message was transmitted from the Arecibo Observatory during the commemoration ceremony on November 16, 1974 and consisted of a series of binary digits that represented a pictorial message intended to convey basic information about humanity and our understanding of the universe. The message contained the following information:

1. the numbers 1 to 10;
2. the atomic numbers of the elements that make up human DNA;
3. formulae of chemical compounds that make up DNA nucleotides;
4. the approximate count of DNA nucleotides in the human genome and a visual of DNA's double-helix structure;
5. the height of an average American man, an illustrated human figure, and the current human population on Earth;
6. an illustrated solar system diagram, indicating the planet of origin for the message, Earth;
7. an image of Arecibo and the dish's physical diameter.

The message was aimed at the Great Globular Star Cluster in the constellation of Hercules, M13, located approximately 7,700 parsecs away from Earth. Drake had learned from the backlash from the Pioneer plaque figures and put thought and consideration into how to present the human figure on the message:

> [On my message,] the human appeared decidedly masculine, much to my chagrin. But when I tried to sketch a more unisex person with the limited number of bits, it came out looking more like a gorilla than a human. So I left it masculine rather than apelike. I may have been trying to unconsciously to make it resemble me, as though I were beaming myself up Star Trek style. I admit I even noted the human's height as... 5 feet, 9 inches tall – just about my size. What a coincidence![42]

Here again we see how, despite efforts to be inclusive, the resulting message still centered masculine depictions of humanity. This is not a judgment on Drake, who, as message designer, might understandably want to represent himself on the message (just as artists have long created self-portraits to ensure their remembrance). Rather it is another piece of evidence that shows how the demographics of CETI scientists affected the character of the messages they sent and, more importantly, how CETI scientists made choices that transformed their messages in subjective ways.

Just as I implored you to view Poetica Vaginal as a type of abject art, I invite you now to consider Drake's message as a form of art, of self-portraiture. In a grand view of history, the self-portrait is actually a rather new type of art. Art historians generally believe that the first self-portraits originated in the fifteenth century (though I should note that this refers to a narrow definition of western self-portraiture; one could certainly argue that earlier forms of art, perhaps even handprints in cave art, were self-portraits). An early prominent example of self-portrait was the one painted by Albrecht Dürer. Before Dürer, appropriate subjects for art were almost exclusively religious ones, such as images of Jesus Christ or the Madonna. The fifteenth century seems to be a rough estimation of the point at which self-portraits emerged in art. We make self-portraits to communicate something of ourselves to others. It is inherently a type of communication: look at me, this is who I am. We also create portraits to immortalize ourselves, or at least a moment of ourselves – an "I was here" moment.

Dürer's self-portrait, however, went an essential step further: here Dürer does not simply do away with more standard subjects of art, such as depictions of Christ, making himself the subject instead. He also downright metamorphoses himself into a depiction of Christ. In the painting he portrays himself with stylish golden curls, symmetrically arranged. He holds an enigmatic gaze indicating a direct gaze at the viewer, and his dignified hand pose refers to gestures known from early Christian iconography. He is both Albrecht and Christ in one.

Similarly, in Drake's depiction, he universalized himself – he was both "a man" and "mankind." With this view, Drake and Dürer were accomplishing the same thing; they were using the medium of self-portraiture to communicate their identity and transcend individuality, tapping into universal themes and symbols. Just as Dürer merged with the divine figure of Christ, Drake's message transcends his personal identity to represent broader human experiences and aspirations. Both artists used self-portraiture not only to assert their presence and individuality but also to connect with others at a deeper, more universal level. In doing so, they created artworks that serve as timeless reflections of humanity and its place in the cosmos. But was the Arecibo message art alone, or was it a genuine attempt at messaging extraterrestrial intelligence?

There are similarities between Drake's Arecibo message and the Soviet Morse message. First, the Arecibo message was sent as part of a dedication ceremony that celebrated a major upgrade to the telescope, just as the Morse message intended to signal the launch of the new facilities at Yevpatoria. Donald Campbell, a professor of astronomy at Cornell who was also a research associate at Arecibo when the message was sent, called the message "strictly a symbolic event, to show that we could do it," as opposed to a rigorous attempt to make contact with extraterrestrial intelligence.[43] A Cornell press release later noted "the real purpose of the message was to call attention to the tremendous power of the radar transmitter newly installed at Arecibo."[44] Second, due to the impromptu and unplanned nature of the message, the message would never reach its intended audience. Drake either had not factored in (or had not bothered to calculate due to the symbolic nature of the event) the rotation of the Milky Way galaxy. By the time the Arecibo message travels 25,000 light years, M13 will have moved well out of the way of its path. Like the Morse message, the message Arecibo sent to Earth was more important.

Unlike the Soviet Morse message, however, the Arecibo message was not explicitly nationalistic in nature, but made an attempt at presenting a universal human message (though, as Drake pointed out in his musings on the masculine figure, an imperfect one). By 1974, tensions between the Soviet Union and the United States in the space realm were at historic lows – in fact, just one year after the Arecibo message was sent the Apollo-Soyuz mission would lead to the symbolic handshake between astronauts and cosmonauts as they listened – and this is true – to "Why Can't We Be Friends?" by the American funk and soul band, War, aboard the docked spacecrafts. The Arecibo message marked a shift in CETI messaging, namely toward conjuring that universalizing cosmic mirror.

Still, although the CETI mission aimed to make contact with extraterrestrial intelligence, it is telling that during the Cold War period the few messages we did send to space were still largely meant to signal meaning to earthlings, not to extraterrestrials. This is further evidence that the CETI project was influenced by the geopolitical climate of the Cold War, where nationalistic and ideological rivalries played a significant role in shaping the priorities and intentions behind our attempts at interstellar communication.

Drake and Sagan were given another attempt at messaging extraterrestrial intelligence through their involvement with NASA. In 1977, NASA had an opportunity to launch two spacecraft, named Voyager 1 and Voyager 2, to explore the outer solar system through visits to Jupiter,

Saturn, Uranus, and Neptune. The opportunity was ideal because the precise locations of the planets – a rare alignment, which only occurs once in 175 years – allowed the engineers to use gravitational slingshots to efficiently accelerate the probes, turning a potentially 30-year journey to Neptune into one of just 12 years. Several months before the launch, Carl Sagan lobbied NASA for another opportunity to attach a message to the probes and received approval. He gathered a small team to design the message; it included himself, Drake, writer Ann Druyan, artist Jon Lomberg, and Linda Salzman-Sagan. The final product would be a set of records, coated in gold to be protected from debris and radiation on their journey through the interplanetary medium. Limited in space, the contents of the record would be able to hold only about 115 images and a short suite of audio recordings with music, spoken greetings, and natural sounds from the Earth. The images on the record depicted the Earth and other planets in our solar system, DNA, and various images of life on Earth, for example sports events, a traffic jam, the Arecibo telescope, a Chinese dinner party, and a school of fish.[45]

Sagan credited his inspiration in designing the Voyager Golden Record back to a childhood visit to the 1939–1940 New York World's Fair – "the World of Tomorrow" – when he was five years old. Walking among the scientific marvels and the displays from civilizations around the world, Sagan claimed to have had an important epiphany, which stemmed from the main message communicated by the fair: "there were other cultures and there would be future times."[46] What if those other cultures included extraterrestrial ones, and in those future times there would be space-faring civilizations? This sparked in Sagan an early interest in communicating with alien cultures, both on Earth and in the universe.

There was a recognition by the Voyager team that the Golden Record was equally, if not mainly, a form of communication with earthlings rather than extraterrestrials. A consultant on the record, Hewlett-Packard executive and engineer Barney Oliver, told Sagan:

> There is only an infinitesimal chance that the plaque will ever be seen by a single extraterrestrial, but it will certainly be seen by billions of terrestrials. Its real function, therefore, is to appeal to and expand the human spirit, and to make contact with extraterrestrial intelligence a welcome expectation of mankind.[47]

Once this truth was recognized, together with the fact that close attention would be paid by earthlings to the contents of the record, understandable concerns emerged that the record would be skewed toward a US-dominant perspective; hence great efforts were made to avoid this bias and to conjure the cosmic mirror once more. Yet the Golden Record team often found itself inexplicably dealing with too many Earth-based problems in a message that purported to concern extraterrestrials.

Sagan decided to include in the record a sample of terrestrial languages and determined that the best thing to do would be to collect audio recordings of simple greetings, such as "Hello." His initial thought was to visit the UN headquarters in New York City, suggesting that a delegate from each member nation stop by the sound studio to give their language's version of "Hello." In addition to national and ethnic diversity, Sagan also hoped to get a roughly balanced gender representation in the greetings.

Unfortunately, he discovered that "virtually all the chiefs of delegations were male, and it was unlikely that they would delegate the privilege of saying 'Hello' to the stars to anyone else."[48] Once again, the gender imbalance in our world bubbled to the surface. It had not occurred to Sagan that this would be a problem until he was forced to face the issue head on. The new realization presented a problem for more reasons than one. It left a burning question in the design of the record: should the team represent the world as it truly was, with its gender imbalance in leadership that resulted from a long legacy of patriarchy, as well as with other forms of oppression such as racism and war? Lomberg, who was tasked with designing the image collection, noted that the team decided against this "truthful" representation of humanity, opting instead for a "best foot forward" approach. There were concerns that depictions of war or nuclear bombs could be interpreted as hostile and threatening to an extraterrestrial civilization. Thus the record was devoid of images of violence, colonialism, slavery, and other human ills.

But even a best foot forward approach presented problems, because humans on Earth could not agree on what made a good first impression. For example, in a collection of essays on the international music content of the record, Sagan recounted a story in which the US team selected "The Young Peddler" as its principal example of Russian folk music. The song, which predated the Soviet Union, had lyrics that told the story of a salesman who interacted with a young woman as they haggled and debated over the price and quality of the goods he was attempting to sell. The song used the argument over goods as a metaphor for romantic courtship and marriage. The Soviet Union was unhappy with this representation of Russian music by a single song that honored a capitalistic transaction, which undermined the country's ideological stance that human society would eventually move to communism. Sagan wrote to an unnamed Soviet colleague asking for a better suggestion. The request was taken seriously and debated thoroughly within the USSR Academy of Sciences, which eventually recommended "Moscow Nights," a popular Soviet song with simple, descriptive lyrics about an evening in Moscow. Unfortunately the response with this choice arrived too late for incorporation into the record; in an attempt to include a song that aligned with Soviet ideals, Sagan had selected instead "Tchakrulo," a Soviet Georgian song about revolt against a tyrannical landlord. Clearly the introspection caused by crafting messages to aliens stirred passions and disagreements that stemmed from conflict and inequity on Earth.

Malevolent or Hungry

As already evidenced by the Morse message, the Pioneer plaque, and the Golden Record, the politics of messaging extraterrestrial intelligence was rife with earthly tensions. In the mid twentieth century, the world was undergoing what is sometimes called "planetary consciousness": an awareness that human beings share (and are responsible for) one planet, a byproduct of globalization.[49] This has already been touched upon as I discussed the concept of the cosmic mirror, which purported to promote a unified global perspective. Planetary consciousness can also be found in the establishment of world government organizations such as the United Nations and the World Health Organization during the mid twentieth century. When it comes to messaging extraterrestrial intelligence, planetary consciousness posed troubling questions. Who should speak on behalf of Earth? Whose perspectives should be represented? And, most importantly, who has the *ability* to speak for Earth (i.e. who has access to the tools and techniques required to send messages). Furthermore, some worried that establishing contact with extraterrestrial intelligence could be dangerous. Shortly after Kardashev published his paper on the scale of civilizations, the *New York Times* printed an article written by the biochemist and author Isaac Asimov.[50] There Asimov took seriously Kardashev's suggestion that CTA-21 and CTA-102 might be evidence of supercivilizations that, Asimov surmised, could be "several thousands of years ahead of us" in technological evolution.[51] Asimov acknowledged that some people may have concerns that messaging extraterrestrial intelligence would bring unwanted attention: "Even we ourselves, so little removed from the Nazi horrors . . . Are supercivilizations to be less decent than our imperfect selves?"[52]

Asimov was not the only party raising concerns. Although Drake had learned from the public backlash against the Pioneer plaque, he did not entirely escape the controversy around the Arecibo message. While his masculine self-portrait largely escaped the public's ire, the act of sending a radio message itself produced considerable disquiet. In his autobiographical writings, Drake claimed that the Arecibo message "provoked a serious outburst from Sir Martin Ryle," the British Astronomer Royal, whom Drake claimed sent the president of the International Astronomical Union (IAU) a letter "full of outrage and anxiety," arguing that "we had exposed humanity as a whole to extraterrestrials of possibly evil intent, who might now prey on us."[53] In a collection of writings on the design of the Voyage Record, Drake claimed that Ryle was concerned: "for all we know, any creatures out there were malevolent or hungry, and once they knew of us, they might come to attack or eat us." Ryle's alleged words have become somewhat infamous in the SETI community, especially his remarks about extraterrestrials being "malevolent and hungry." Such formulations have been quoted and attributed to Ryle in dozens, if not hundreds, of publications such as the *New York Times*[54] and the *Guardian*[55] and in many books[56] on C/SETI topics.

But these words are misattributed. Ryle's letter to the IAU was mischaracterized by Drake, perhaps to undermine the legitimacy of those who raised concerns about messaging extraterrestrial intelligence. The actual letter sent by Ryle in response to the Arecibo message is housed in the Churchill Archives Centre at Churchill College, Cambridge. It was sent a couple of years after the message was sent, in 1976. Ryle was writing to Bernard Lovell, then vice-president of the International Astronomical Union's Executive Board, and began the letter by stating that he first became aware of Drake's message via a BBC program on extraterrestrial life – and not immediately after the message was sent, as Drake suggested.

The letter was not hysterical, but rather raised questions about political implications and the potential harm that could arise from revealing our existence to more advanced civilizations, though being preyed upon was not a listed concern. Instead, Ryle pointed to the history colonialism on Earth and to "Earth as a useful place for colonization/mineral extraction." As I noted earlier, during the early Cold War period, between 1945 and 1960, dozens of states in Asia and Africa gained independence from their European colonizers, especially Britain. Perhaps as a British citizen who lived through the decolonization of the British Empire, Ryle was intimately familiar with the consequences of colonialism and the dangers of resource-hungry, expansionist civilizations. The letter continued by pointing out that, even if colonization was not a concern, Ryle believed that international consultation was necessary before undertaking such transmissions, since they would inherently represent the entire Earth. In his view, a responsible and democratic process was required. He was especially suspicious of the United States' taking the lead in messaging: "I do not feel that it is right to leave the decision to one group of one nation (who have in any case a poor record of taking on responsibility for the Earth, with the Needles and Starfish programs)."[57]

The two programs Ryle derisively references were US Cold War military space programs. The first, Project Needles, later renamed Project West Ford, was an endeavor by the US Air Force and the Department of Defense to safeguard the nation's long-range communications, in response to the escalating tensions with the Soviet Union. Their goal was to create the largest radio antenna ever built by launching approximately half a billion extremely thin copper wires into orbit. However, as Project West Ford advanced in its development, concerns were raised by radio astronomers, especially those at NRAO, regarding the potentially adverse consequences of this metallic cloud on their ability to observe. Furthermore, the initial payloads were highly criticized by the international community. Shortly after the first launch, the Soviet newspaper *Pravda* apparently published an article that condemned Project West Ford under the title "USA Dirties Space."[58] Project West Ford diminished the reputation of the United States as a responsible actor in space, especially in eyes of the international radio astronomy community.[59]

The second project that Ryle listed as evidence that the United States could not be trusted to engage responsibly with space was "Starfish." He was referring here to Starfish Prime, a nuclear bomb detonated in space by the US Air Force, approximately 248 miles above the Pacific Ocean, in July 1962. According to space historian David Portree, the purpose of the Starfish Prime program was to examine the potential of nuclear explosions in low-Earth orbit for enhancing and extending the Van Allen radiation belts that encircled the Earth. The goal was to create a barrier capable of rendering ineffective the Soviet intercontinental missiles launched toward the United States. This is yet another example of the militarization of outer space in the early Cold War period.[60]

Drake claimed that Ryle "wanted the IAU Executive Committee to condemn the [Arecibo message] in an official resolution."[61] But Ryle's actual suggestion was mellow and sensible. He asked Lovell:

> As Vice-President of IAU you might feel that this is something which should be discussed at [at the next IAU meeting]. But the principle goes much wider, and in my view no such programmes of transmission should be undertaken without an international discussion involving much wider participation than astronomers alone.[62]

Thus we can see a large disconnect between the events that took place and our historical memory. How do disconnects like this happen? Ryle's concerns were reasonable and in line with the larger CETI goals of establishing international communication. Yet he was characterized as holding an extreme position, to a lasting effect. Ryle is still regularly quoted in the SETI literature as a major party opposed to messaging extraterrestrial intelligence, and the words put in his mouth, "malevolent and hungry," live on.

There is an effect of negativity bias, where human beings sometimes distort, exaggerate, or weigh too heavily the criticism they receive. Perhaps Ryle's mild criticisms felt disproportionately harsh to Drake, especially as they came from a Nobel laureate and an Astronomer Royal. But it is also possible that Drake was trying to undermine Ryle's critique and frame it as unreasonable. As a small new field, not always taken seriously in astronomy, CETI was vulnerable and needed scaffolding, even against reasonable concerns. CETI actors were internationalist, but were not always willing to work with the international community if that obstructed their goals.

As we have seen, messages to extraterrestrial intelligence often took a national character during the Cold War, despite their aims at universalism. And even when attempting to create a universal representation, CETI highlighted nationalism, paternalism, and other earthly tensions. In this sense, the cosmic mirror did not create the universalizing "reflection" of our world that its proponents hoped it would, but a "refraction." Rather than looking inward and reflecting our culture outwards, CETI created a slightly distorted vision of our world, one subject to changes that depend on the medium it passes through. In doing so, CETI tells us something about our world: just as we learn about the properties (density, temperature, etc.) of the materials that light refracts through from these refractions, so too do we learn about our world by examining its distortions – as in a cosmic prism.

6

The Cosmic Prism
CETI and Existential Threat

Shortly after the end of World War II, in 1950, Enrico Fermi, a physicist sometimes referred to as the "architect of the atomic bomb,"[1] visited the Los Alamos National Laboratory in New Mexico. One afternoon during this visit, over lunch, the conversation drifted toward the extraterrestrial life debate. Fermi and his companions were discussing UFOs, perhaps prompted by the 1947 flying disc craze, which had shaken the country just a few years earlier, thanks to a rash of UFO spottings that included the scandalous Roswell incident. As the scientists began to tackle the probability of a universe full of life, Fermi asked a now infamous question: "But where is everybody?"[2]

This question about the proliferation of life in the universe was raised against the backdrop of a nuclear laboratory that, only five years earlier, had created the first two atomic bombs and dropped them as an act of war. Fermi's question predated the development of CETI by several years, but it underlines the strong connection between speculation about extraterrestrial civilizations and the future of our own.

The specter of the Cold War has hung over every chapter of this book. This final chapter, however, focuses not so much on science, infrastructure, or communication methods as on something rather different. This chapter deals with emotion.

We have already seen how CETI scientists' vision of extraterrestrial civilizations was tied up with their own projections for the future of human civilization. Kathryn Denning has noted that, when CETI scientists engaged in speculation on the alien future, this "was itself an act of hope."[3] CETI scientists reached across the Iron Curtain in the hope of using CETI to establish friendlier relations on Earth. But here I want to shift our attention to hope's underbelly: fear. Hope and fear are intimately connected. Psychologist C.R. Snyder writes that "they are both motivators, one towards something and one away from something."⁴ Although CETI scientists were motivated by the hope that came from its internationalist ideals, they were motivated just as much – if not even more – by fear of the collapse of Earth's civilizations.

The ideals of the cosmic mirror were not exclusive to CETI and can be found in many elements of the space sciences during the Cold War. A publication by NASA argued that space sciences are driven by both sectarian international rivalry and an innately human desire to explore, which according to the authors leads to "national embodiments of a universal human drive."⁵ This "universal" and internationalist rhetoric often obscured hostile nationalistic aspects of spaceflight. Internationalist quotes from astronauts make headlines – such as this one from Apollo astronaut Edgar Mitchell:

> In outer space you develop an instant global consciousness, a people orientation, an intense dissatisfaction with the state of the world, and a compulsion to do something about it. From out there on the Moon, international politics look so petty. You want to grab a politician by the scruff of the neck and drag him a quarter of a million miles out and say, "Look at that, you son of a bitch."⁶

Such a statement supports the case for the existence of a cosmic mirror in the space sciences – that reflection on human civilization in its cosmic situatedness inevitably opens internationalist, apolitical perspectives. This perspective is also sometimes referred to as "the overview effect," which describes a cognitive shift in the minds of people who have viewed the Earth from space and gained a sense of interconnectedness and awe. But, as the NASA psychology study showed, this rosy picture sometimes masked the other complex reality of national conflict in space exploration. Astronauts and cosmonauts flying joint missions (such as those on Apollo-Soyuz in 1975) had to balance their nationalistic images (as a symbol of communism or democratic capitalism) with a rhetoric of internationalist scientific collaboration, but also with the natural intercultural discomfort that can arise when you share a small space with a stranger who has different values and habits.⁷ A similar conflict can be found in CETI; while scientists touted an internationalist rhetoric and antinational perspectives, they faced many conflicts that stemmed from national tensions. With this in mind, we can expand on Tarter's view of the cosmic mirror: it highlights not just our shared humanity, fostering internationalism, but also the conflict and tensions within our world.

Seen in this expanded context, the cosmic mirror might be best understood as a form of the sublime, a concept developed by the British philosopher Edmund Burke in a mid eighteenth-century treatise devoted to it.[8] Contemporaneous artists such as J. M. W. Turner captured the sublime by painting scenes of nature such as storms and big waves, which inspired a sense of awe and captured the feeling of being small and powerless when confronted by the great forces of nature (Figure 6.1). The cosmic mirror is often evoked in popular writings on the space sciences, for instance in Carl Sagan's *Pale Blue Dot* (1994), which, like the Romantic sublime, elicited a sense of wonder that stemmed from an understanding that we are small in the face of an endless universe. Sagan wrote about the image of Earth in a photograph taken by the Voyager probe (Figure 6.2). Seeing the small light speck of the Earth surrounded by the depths of interplanetary space, Sagan stated:

> Our posturings, our imagined self-importance, the delusion that we have some privileged position in the Universe, are challenged by this point of pale light. Our planet is a lonely speck in the great enveloping cosmic dark. In our obscurity, in all this vastness, there is no hint that help will come from elsewhere to save us from ourselves.[9]

The description conveys a sense of helplessness, and Sagan uses that feeling to argue for peace and empathy, stating that the humility brought upon by the Voyager image should underscore "our responsibility to deal more kindly with one another, and to preserve and cherish the pale blue dot, the only home we've ever known."[10] Interestingly, Burke declared that "whatever is in any sort terrible or is conversant about terrible objects or operates in a manner analogous to terror, is a source of the sublime," drawing a direct line between wonder and fear.[11] That kind of fear is present, albeit subtly, in Sagan's description of the Pale Blue Dot, with his insistence that we must save ourselves. This combination of fear and hope is a fundamental part of CETI, largely brought about as it was by Cold War anxieties and a reckoning with mortality – not only that of the individual, but of the entire species.

Figure 6.1 Joseph Mallord William Turner, Snow Storm: Steam-Boat off a Harbour's Mouth. Exhibited in 1842. Credit: Wikimedia Commons.

Figure 6.2 The Pale Blue Dot, captured by the Voyager 1 probe. Credit: NASA/JPL-Caltech.

Longevity Anxiety

Underpinning the entire history of CETI is a tension between fear and hope – fear that our civilization might be destroyed and hope that we might escape this fate. This tension is clear even in the earliest CETI project – Drake's Project Ozma. Shortly after its completion, this project had stimulated an interest in CETI in the greater scientific community, and so the Space Sciences Board of the National Academy of Sciences (NAS) decided to support what they called a "quiet meeting" in the autumn of 1961, to discuss the possibility of making radio contact with extraterrestrial intelligence.[12]

Although small, the first US CETI conference was attended by an elite group of scientists from several fields. Among them were the director of NRAO, Otto Struve; the future Nobel prize winner in physics, Charles Townes; Melvin Calvin, Nobel prize winner in chemistry (in fact he learned during the conference that he won the prize); Su Shu Huang, NASA astrophysicist; Philip Morrison, a scientist from Project Manhattan; and John C. Lilly, a biologist who researched dolphin communication.[13] The scientists at the conference drew connections between Lilly's work and their own: the pursuit of communication with the non-human intelligence and with unknown intelligence. As a result, they dubbed themselves "the Order of the Dolphin," a semi-secret CETI society.[14] Sagan was the youngest scientist in attendance; he was perhaps invited mainly for his connections with Melvin Calvin, who had written a letter of recommendation that helped him acquire a postdoctoral position at UC Berkeley.[15]

The Space Sciences Board had outlined three major topics to be addressed at the conference. First, they wanted the attendees to consider the "estimates of limiting values for the probability of existence of planets on which civilized life is likely to have evolved."[16] Second, the attendees should determine whether this number was sufficiently large and should decide whether current search methods were useful enough to make the search worth pursuing. And, finally, the conference attendees were to make recommendations to the Board for further study.[17] To address the issue of feasibility, Drake devised an equation that became subsequently known as "the Drake equation." The equation was simple: a product of factors, the values of which were largely unknown at that time. The

Drake equation was represented as "$N = R^* \cdot f_p \cdot n_e \cdot f_l \cdot f_i \cdot f_c \cdot L$," with each variable defined as follows:

N = The number of civilizations in the Milky Way galaxy whose electromagnetic emissions are detectable.
R^* = The rate of formation of stars suitable for the development of intelligent life.
f_p = The fraction of those stars with planetary systems.
n_e = The number of planets, per a planetary system associated with one star, with an environment suitable for life.
f_l = The fraction of suitable planets on which life actually appears.
f_i = The fraction of life bearing planets on which intelligent life emerges.
f_c = The fraction of civilizations that develop a technology that releases detectable signs of their existence into space.
L = The length of time such civilizations release detectable signals into space.[18]

Since the equation could not be reasonably calculated with their knowledge at the time, it was designed to function as a tool that the attendees of the conference could use to frame their estimates of extraterrestrial intelligence in the galaxy.

The last variable, L, has been the focus of many studies in technosignature literature, with estimates ranging from incredibly optimistic (upwards of millions of civilizations) to deeply bleak (only humans).[19] As time has progressed and fields of study such as exoplanetary science have developed, many of the Drake equation variables have begun to fill in, but the answer to L has remained elusive and highly relevant. Kathryn Denning has argued that L is the variable in which "we are exceptionally emotionally invested," because "our estimates of L are intertwined with our forecasts for our own civilization's end."[20] For that reason, Drake's inclusion of L in the equation is particularly revealing of the cultural climate in which CETI developed; after all, almost exactly one year after the first CETI conference in Green Bank, the world would experience its first nuclear stand-off – the Cuban missile crisis. The question of L was inextricably tied to concerns about the role of technology in the destruction of life and civilization, concerns that were extremely pressing at the height of the Cold War.

The rise of détente[21] culminated in the creation of scientific exchange programs that aimed to alleviate some of the tensions between the United States and the Soviet Union. One such formal agreement between the two countries was an "Agreement on Exchange of Scientists between the National Academy of Sciences of the USA and the Academy of Sciences of the USSR" in 1970 and 1971, which would support the exchange of individual scientists from each country for scientific, diplomacy, and information exchange visits.[22] Nikolai Kardashev became one of the scientists involved in these exchanges. He was invited to visit the National Radio Astronomy Observatory in 1970, after a joint US–USSR radio astronomy experiment. According to Soviet CETI scientist Lev Gindilis, at some point during this trip Kardashev met Sagan, and the two decided that the agreement between the two Academies should be used support CETI, given its peace-making capacity. Sagan subsequently approached the US National Academy of Sciences to sponsor a joint US–USSR CETI conference, arguing that such a conference would support the goals of the NAS agreement, which stated that "the National Academy of Sciences of the USA and the Academy of Sciences of the Soviet Union agree on the desirability of conducting in the USA and the Soviet Union jointly sponsored symposia on important scientific problems."[23] In his proposal, Sagan argued that "contact" between the United States and the Soviet Union on the issue of CETI might lead to the development of scientific collaboration between the two nations:

> The time seems ripe to examine seriously these important issues. Even if extraterrestrial intelligence proves to be so rare as to be indetectable in the near future, the potential contribution of such meeting to many interdisciplinary scientific questions seems to be very great. It is possible that this contact will lead to cooperative research programs by scientists of the US and USSR.[24]

The conference was approved and supported by the Academies in both nations. A joint US–USSR CETI conference was swiftly organized, hosted by the Soviet Byurakan Observatory in Armenia; Sagan would chair the US side and Shklovsky would chair the Soviet side.[25] I have already briefly referred to the US–USSR CETI conference as a means of addressing the interdisciplinary conflicts within CETI and of highlighting the determinist perspectives of the scientific attendees.

Figures 6.3–6.6 *Top Left:* Nikolai Kardashev presenting a paper at the US–USSR CETI conference. *Top Right:* Shklovsky "making a point with some vigour to Philip Morrison," with Sagan listening from behind. *Bottom Left:* Sagan and Shklovsky, who is described as being "unimpressed by an argument" made by Sagan. *Bottom Right:* Frank at a desk in the Byurakan Observatory, Soviet Union. Credit: Photos by Phyllis Morrison, in Sagan, Carl, ed. *Communication with Extraterrestrial Intelligence (CETI)*. Cambridge: MIT Press, 1973. Reprinted courtesy of the MIT Press.

Here I will focus on how Cold War consciousness influenced the nature of the discussion and subjects raised.[26] The Cold War shaped not only the infrastructures, rhetoric, and collaborations of CETI but also the mentalities of CETI scientists. The tensions in the field were not lost on its practitioners, and as a result their cognizance of Cold War anxieties culminated in a great focus on existential issues regarding life in the universe (see Figures 6.3–6.6).

There were 54 participants in attendance at the conference, with ten sessions dedicated to problems such as "the origin of life," "the evolution of intelligence," and "the possible consequences of establishing contact with extraterrestrial civilizations."[27] Among the American delegation were most of the US CETI pioneers, including Frank Drake, Philip Morrison, Freeman Dyson,[28] and Kenneth Kellermann. The Soviet delegation included many CETI scientists whom we have already met, such as Lev Gindilis, Nikolai Kardashev, and Yuri Pariiskii. Although the conference was officially a joint US–USSR conference, there were four participants from other countries: Britain, Canada, Hungary, and Czechoslovakia sent one representative each.[29] Interestingly, Britain's representative was Sir Francis Crick, Nobel Prize winner and co-discoverer (along with Rosalind Franklin and James Watson) of the helical structure of the DNA molecule. Later in his career, Crick had become preoccupied with the origins of life and promoted a theory called "directed panspermia," which argued that life might have been deliberately "seeded" throughout the universe by intelligent extraterrestrials, an argument also considered by Shklovsky and Sagan in *Intelligent Life in the Universe*.[30]

The American delegation was particularly interested in the Soviet search effort, which they believed was more readily embraced in the Soviet Union than it was in the United States. In fact, in his NAS proposal, Sagan specifically argued that one reason why the conference was a necessity was that, "through a state commission in the Soviet Union for the study of communication with extraterrestrial intelligence, there exists a more active theoretical and observational program in this area than exists at the present time in the US."[31] As a result of the centralized funding systems in the Soviet Union, which were less competitively based than in the United States, Soviet institutions received generally the same number of funds each year, and senior scientists determined how those funds were allocated. This Soviet organization of science was created as a deliberate foil to western science; as historian of Soviet science Loren Graham has noted, the Soviets believed that major flaws in western science stemmed from "inefficiency due to competition among secretive independent industries, lack of centralized planning, and inadequate financial support from the government," and consequently created a system in which science funding was centralized and government-supported.[32] Therefore, if a science had the backing of key governmental or institutional figures, it would receive government funds. If a highly respected scientist who headed an institution (such as Shklovsky) decided to promote CETI, a search could be arranged, as was the case with the CTA-102 observations. The US scientific community was decidedly more hostile to CETI observational efforts, and by 1970 Drake's Project Ozma remained the only radio CETI observation conducted in the United States a decade after it took place.³³ Drake believed there was a strong political motivation behind the Soviets' receiving state support for CETI. He believed that Soviet authorities "correctly perceived the search enterprise as an area where Soviets could compete with and possibly excel over American efforts" and that, unlike the United States,

> Soviet scientists at big institutions had money to spend on search efforts, and met no opposition when they opted to build special equipment or staff projects devoted to detecting alien civilizations. And while I wish we American astronomers had enjoyed a warmer reception on this account, I believe that the Soviet acceptance was not entirely benign. Indeed, it had little to do, in my opinion, with wide regard for the search enterprise itself, but . . . there was also a political motive behind the governmental support for these activities.³⁴

Drake's suggestion neglects to consider the complexities of Soviet funding systems, of course. As Benjamin Peters argues in *How Not to Network a Nation*, there was intense interministerial competition over scientific and technical funding in the Soviet Union, and this competition was mired in politics and personality battles.³⁵ Nevertheless, it is true that individual Soviet scientists had more autonomy in choosing which projects to support, so long as there was institutional backing. In the United States, on the other hand, a CETI project often had to piggyback off what was considered to be a more serious scientific one. Drake, in conducting Project Ozma, had to make sure the tools he developed for his search, such as the radiometer, could be used for more traditional scientific purposes, like searching for evidence of the Zeeman effect.³⁶

* * *

Despite these subtle undertones of jealousy or competition (Shklovsky described the atmosphere: "hot discussions flared sporadically"),³⁷ the 1971 conference was laden with internationalist rhetoric, further cementing the parallels – established by Sagan in the conference proposal – between contact with extraterrestrial aliens and contact with terrestrial aliens. In fact, in the conference opening remarks, Viktor Ambartsumian, a Soviet astronomer and the director of the Byurakan Observatory, paraphrased a conversation he had with Shklovsky:

> Professor Shklovsky was right when he said to me that before we are able to solve the problem of communicating with extraterrestrial civilizations, it might be a good thing for there to be communication on the subject among nations, and that is precisely the purpose of our conference.[38]

The conference proved to be quite lively, with participants who agreed and disagreed on many details regarding the search for and communication with extraterrestrial intelligence, including the likelihood of successful communication. Some advocated the use of the "universal" language of mathematics, while others argued that, if human beings could not meaningfully communicate with cetaceans (another arguably intelligent species on Earth), there could be no hope for communicating with extraterrestrial intelligence, which would be far more alien than dolphins and porpoises. One thing all participants seemed to agree on, however, was the imagined impact of CETI science on "the future development of mankind," a sentiment oft repeated throughout the conference.[39] The conference proceedings clearly illustrated how the CETI scientists considered their field; they saw it as a great unifier, a science that could operate beyond politics and bring about universal peace. The conference participants felt that their "unanimity of purpose" provided them with "the courage to plot together for the future of all humankind."[40] Their faith in the peace-making capabilities of CETI were ambitious to the point of absurd; Drake mentioned that there was discussion of constructing a CETI radio telescope that straddled the Israel–Egyptian border for the purpose of searching for extraterrestrial intelligence "while promoting peace in the Middle East."[41] Given this enormous potential in the minds of CETI scientists, the conference resulted in the formation of an international working group meant to "coordinate national programs of research and to promote progress in this field," whose members included Frank Drake, Nikolai Kardashev, Philip Morrison, Bernard Oliver, Rudolph Pešek, Carl Sagan, Iosif Shklovsky, Vsevolod Troitskii, and G. M. Tovmasyan.[42]

The 1971 US–USSR CETI conference had a far-reaching impact on the scientists in attendance. Drake later wrote about the excitement it generated; during the conference, the scientists involved "laid the groundwork for a number of new searches by both American and Soviet radio observatories," and also engaged in discussions regarding the role of CETI in peaceful international collaboration.[43] The continued emphasis on peacemaking, cooperation, and internationalism masked a significant undercurrent: many of the participants, including Shklovsky and Sagan, were fully aware of the numerous barriers and dangers of the Cold War atmosphere in which they were operating, and this awareness shaped the subjects discussed at the conference. This manifested largely in a focus on L – longevity – in the Drake equation, as conference attendees worried about whether technologically advanced civilizations might bring about their own destruction. Among those invited to the Byurakan conference was a historian, William McNeill, who had achieved notoriety for his earlier publication of *Rise of the West* (1963). The presence of a historian whose expertise was in the rise and fall of civilizations reflected the pressing concern that L raised in the minds of CETI researchers. Shklovsky led a session titled "The Lifetimes of Technical Civilizations," in which the discussion focused on the threats of "nuclear destruction, pollution, ecological disruption, overpopulation, and exhaustion of natural resources."[44]

Yet discussions around L did not constitute the only aspect of the conference that highlighted Cold War anxieties. James Elliot, an astronomer then affiliated with the Laboratory for Planetary Studies at Cornell University, presented a talk titled "X-Ray Pulses for Interstellar Communication."[45] The title only hints at the content which began by noting that X-rays might be a useful medium for transmitting an "announcement message" in CETI and pointing out that humans have "already used X-rays to send out signals" to the cosmos.[46] How was this so? Because, as Elliot explained, "these signals were sent in the course of several nuclear explosions" conducted by the United States and the Soviet Union.[47] Elliot argued that, if the United States and the Soviet Union "pooled their nuclear stockpiles to produce a single large explosion," pausing to add "far from Earth!," the X-rays generated from such an explosion could "be detected at a considerable distance." Elliot then spent some time speculating what the existing nuclear arsenal might be and whether it was sufficient for such a message.[48]

173

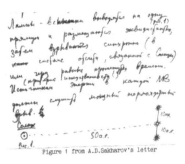

Figure 6.7 An illustration from Soviet nuclear physicist Andrei Sakharov's response to a questionnaire sent out to scientists for the 1971 US–USSR CETI conference. Credit: http://lnfm1.sai.msu.ru/SETI/eng/articles/sakharov.html.

Independently, Andrei Sakharov, an outstanding physicist and human rights activist and one of the leading figures in the Soviet thermonuclear project, also proposed a CETI system that exploited thermonuclear explosion. A friend of Shklovsky's, he formulated his proposal as an answer to the questionnaire for the First American–Soviet CETI Conference of 1971 in Byurakan (Figure 6.7).[49] In his proposal, Sakharov suggested that "flash lamps" made from "powerful thermonuclear explosions" should be placed in our solar system every 10–20 years, evenly spaced along a straight line.[50] These lamps should be exploded simultaneously or at regular intervals, creating short flashes of visible light and microwaves that could be detected by extraterrestrial intelligence. The proposals of using thermonuclear weapons from Sakharov and Elliot implicitly suggested that Soviet–American CETI cooperation could kill two birds with one stone: nuclear disarmament and extraterrestrial communication.

* * *

The Byurakan meeting prompted a number of new searches by both US and USSR radio observatories, including what Drake described as an "orbital radio" (i.e. a space-based telescope) "for the pursuit of the elusive signals" (perhaps a reference to what would later become RadioAstron). Drake hailed the accomplishments of the conference, noting that all the speakers' plans and discussions were conducted "under the shadow of the Bay of Pigs, the Cuban missile crisis" and the KBG's intense surveillance of the Soviet scientists and their visitors.[51]

Up until this point we've seen how Cold War anxieties hung over the meeting. But, as Drake pointed out, the conference attendees had to deal with consequences of the Cold War that went beyond existential concerns. In recollecting the meeting, Minsky noted that the US attendants were constantly supervised by so-called 'interpreters'.[52] On one occasion, Minsky and Crick decided that they had become tired of being supervised, so they secretly left through the rear exit of the Academy of Sciences Hotel, to avoid being followed on a walk around the city. Upon return, Minsky recalled how they found their "interpreter" in tears. She told them that "she feared she'd be punished severely if she 'lost' us again."[53] The parallels between the task of deciding how best to communicate with aliens and that of visiting a technologically advanced foreign nation to hold discussions via translators and with intimidating interpreters were not lost on the scientists in attendance at the meeting. In fact the scientists embraced this view, Drake noting that, when the American delegation left, they felt that they had "gotten a hint of what life on another world was like."[54] Although friendships were formed between American and Soviet CETI scientists as a result of the conference, the challenges of conducting transnational CETI would remain in the following years.

In 1973, Yuri Pariiskii, then the director of radio telescope operations at the Pulkovo Observatory in Leningrad, attended a meeting of the International Astronomical Union in Sydney, Australia. According to Drake (who was also in attendance), Pariiskii approached both him and Sagan to inform them he had detected extraterrestrial signals that were "broad-band, like noise," detected "for a few hours" before disappearing, and recurred every day over the course of several months.[55] Pariiskii noted that the signals seemed intelligent because they were encoded in 1, 2, 7, and 9 pulses. He had already approached Soviet military authorities about his detection and was told that the signals were not coming from any known Soviet or American satellites. Pariiskii then told Sagan and Drake that he was cautious about going public with the news because of the fallout from what he called the CTA-102 "fiasco."[56]

But his wise reticence had presented a problem: up until that point, he had not been able to consult with his American colleagues on this detection because, as Drake noted, "it was dangerous in those days for a Soviet scientist to divulge research data to an American through the mail, as censors read all the letters."[57] This concern was not without merit, as we have already learned. So Pariiskii had to wait for an opportunity to speak with an American colleague in person, and he told Drake that their collective CETI goals "transcended questions of national identity."[58] This was both true and untrue. On the one hand, CETI goals transcended national ones for the scientists engaged in them, but on the other hand these scientists' findings had to be conveyed in person, because those who were engaged in monitoring the communication of intraterrestrial intelligence would be too likely to intercept them. But if the information had been published, it might only have revealed that the scientists who searched for extraterrestrials did not have the imagination (or the classified knowledge) required to recognize evidence of Earth-based intelligence.

Having already experienced several cases of military and intelligence interference in his scientific career, Sagan was skeptical of the signal, and after several weeks of inquiry was able to confirm the existence of an American intelligence satellite called Big Bird, which had an orbital pattern that aligned with the pattern of Pariiskii's signal. Drake wryly recalled: "Big Bird did not represent extraterrestrial intelligence, but it was in the business of intelligence-*gathering* from an extraterrestrial vantage point."[59]

The Common Enemy

Clearly CETI scientists were forced to confront two incompatible ideas: that CETI promoted internationalism and peace, but also that the practice of CETI highlighted the highly nationalist and hostile world they lived in. It is no wonder, then, that the final variable of the Drake equation became such a point of interest in the community. Sagan had fewer constraints than Shklovsky in speaking out against nuclear weaponry (this had led in part Sakharov to exile), so this preoccupation with the destruction of civilizations prompted him to dedicate much effort to antinuclear activism. In 1983 he published in *Foreign Affairs* an essay where he argued that, if the United States and the Soviet Union did not reverse their arms race, there would be "a real danger of the extinction of humanity."[60] Sagan's approach to nuclear disarmament differed from that of politicians such as President Ronald Reagan, who in that same year had described the Cold War as a "struggle between right and wrong and good and evil," with the United States on the side of good. Instead of relying on dichotomies, Sagan used CETI internationalism in his antinuclear rhetoric.

For example, in 1985 President Reagan (a known fan of science fiction) once asked General Secretary Gorbachev if the Soviet Union would rescue the United States in the case of an extraterrestrial invasion. Gorbachev replied, "no doubt about it."[61] This odd conversation inspired the premise for an essay that Sagan penned in 1988, titled "The Common Enemy." In the essay, Sagan begins:

> If only, said the American President to the Soviet General Secretary, extraterrestrials were about to invade – then our two countries could unite against the common enemy... An alien invasion is, of course, unlikely. But there *is* a common enemy – in fact, a range of common enemies, some of unprecedented menace, each unique to our time. They derive from our growing technological powers and from our reluctance to forgo perceived short-term advantages for the longer-term well-being of our species.[62]

Sagan's essay took an antinationalist stance by appealing to the common humanity shared by citizens of the United States and of the Soviet Union, critiquing both nations, and noting that "each side has a long list of deeply resented abuses committed by the other."[63] The essay is driven by concern over the longevity of human civilization, strongly implying that without cooperation the weapons and hostility created by the Cold War would end civilization. Throughout the essay, extraterrestrials were a common theme. Sagan concluded with a call to action, relying on the sense of common humanity and internationalism he had tried to instill (or awaken) in that piece:

> Is it possible that *we* – *we* Americans, *we* Soviets, *we* humans – are at last coming to our senses and beginning to work together on behalf of the species and the planet? Nothing is promised. History has placed this burden on our shoulders.[64]

Sagan's article was published jointly in the American magazine *Parade* and in the Soviet magazine *Ogonyok*. Fascinatingly, the article was censored when it was published in *Ogonyok*. This should not be surprising to the reader of this book, given the time we have spent exploring the Soviet censorship programs. What is interesting, however, is what was censored and why. Much of what was removed was criticism of the Soviet government and Soviet history, while Sagan's criticism of the US government and US history remained (which had the effect of watering down his message). The most significant removal, however, was of the line "nothing is promised" at the conclusion of Sagan's essay. As I have already pointed out (p. 00), one of the tenets of Soviet Marxist philosophy was the belief that the eventual human progression toward communism was "foreordained"; it was a unilineal progression that was ultimately projected on other areas of Soviet life as well.[65] Hence it was almost a sacrilege to say "nothing is promised." Once again, it became clear that internationalist CETI rhetoric often found itself uncomfortably confronted with nationalist tension.

It is important to note that Sagan was not the only CETI scientist to become an antinuclear activist; Philip Morrison, recognized for his co-authorship of the first CETI paper in 1959, had previously served as a leader on the Manhattan Project and oversaw the assembly of the bomb that would detonate above the city of Nagasaki. After viewing the devastation in Japan as part of the Manhattan Project's survey team, Morrison became an adamant antinuclear activist and founded the Federation of American Scientists and the Institute for Defense and Disarmament Studies. After the war he dedicated much of his life to CETI and chaired the early NASA SETI workshops and studies.[66] He argued that CETI's cosmic mirror was more like a crystal ball – something that shows us our own future. He claimed that CETI should be called the "archaeology of the future," explaining:

> Archaeology of the past is very interesting because it tells us what we once were. But archaeology of the future is the study of what we're going to become, what we have a chance to become . . . it's a missing element in our understanding of the universe which tells us what our future is like, and what our place in the universe is. If there's nobody else out there, that's also quite important to know.[67]

The implication of Morrison's statement is that our projections of the future through CETI offer valuable information about our own destiny and position in the universe. Relying on the determinism common in CETI, Morrison argues that extraterrestrial futures tell us about our own. And, critically, he notes that an empty universe might indicate a propensity of intelligent civilizations to destroy themselves. As a CETI scientist, Morrison became a champion of antinuclear proliferation, arguing that, if the United States and the Soviet Union did not remove their possession of nuclear weapons, "sooner or later destruction will come to the cities of the earth."[68] At least in part, the search for intelligent life on other worlds had prompted its practitioners to fight to preserve civilization on Earth.

A Warning about Danger

The previous chapter dealt with messages CETI scientists designed for extraterrestrials, but argued that many (if not all) of these messages were, in some capacity, really for a terrestrial audience. But what about messages CETI scientists designed explicitly for human beings, without the extraterrestrial middleman?

In 1979, just a few years after the US–USSR CETI conference, Congress authorized the establishment of a facility for depositing the country's nuclear waste. Part of the Department of Energy (DOE), the Waste Isolation Pilot Plant (WIPP) facility would be located in southeast New Mexico. The DOE notes that WIPP was "critical to the cleanup of Cold War nuclear production sites" and was built inside a salt basin. In 1957 the National Academy of Sciences suggested using salt for the disposal of radioactive waste. The rationale behind this recommendation was that salt exhibited a plastic deformation known as "salt creep" in the mining industry. This process would gradually fill and seal any voids or gaps formed during mining activities, effectively containing the waste within and around them.[69] The site was licensed to store transuranic nuclear waste for 10,000 years.

Because of its long half-life and the potential for the site to remain dangerous for thousands of years, the DOE created a project, the WIPP, tasked with sending a message to future humans to warn them about the potential danger of the site. This project presented a challenge, however, because it was difficult to know how future human civilizations would interpret various languages or symbols. Therefore the DOE hired a committee of experts with backgrounds in history, economics, linguistics, law, agriculture, geography, and more.

Critically, the DOE was interested in the perspective of the CETI community and invited participants such as Carl Sagan and Frank Drake, as well as Jon Lomberg (an artist who assisted Sagan with designing the Voyager Golden Record) and Woodruff Sullivan (a radio astronomer, historian, and CETI enthusiast). The WIPP's "Expert Judgement" report highlighted the reasoning behind including members of the CETI community by pointing out their experience with "previous efforts to think broadly about communication in terms of using radio signals or sending a satellite into space to communicate over long time periods with unknown beings."[70]

The WIPP consisted of two development panels, the Markers and the Futures. The Futures Panel was tasked with envisioning future scenarios that might result in the radioactive waste being uncovered prematurely. This included postulations on possible environmental disasters and future scientific or archaeological expeditions. The Markers Panel was tasked with determining what types of marker or messages might be constructed to discourage future disruption of the site. The Markers Panel was split into two teams, Team A and Team B. Each team was given the same set of questions to address during its deliberations. Their members were asked to consider not only what type of marker might be constructed, but also what the impact the marker could have in deterring inadvertent intrusion, how long it might persist and be recognizable or correctly interpreted, and how likely it was that the message, if interpreted correctly, would be successful in deterring the future civilization.

A number of suggestions were made by the Markers group. The group envisioned various levels of communication. Levels One and Two would convey rudimentary and cautionary information, warning anyone who might come across the site that something human-made and dangerous was there. The panel proposed creating a hostile artificial landscape above the site, such as a field of giant spikes bursting from the ground, to indicate danger below (Figure 6.8). It also crafted basic written messages, designed to be displayed in many languages. One proposed message read:

> This place is a message . . . and part of a system of messages . . . pay attention to it! Sending this message was important to us. We considered ourselves to be

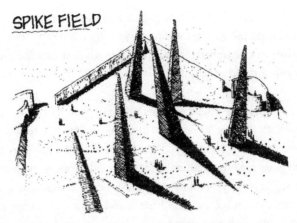

Figure 6.8 Spike Field, view 1 (concept and art by Michael Brill). The Spike Field was one proposed site marker for WIPP. The designers hoped that the spikes might indicate non-verbally that "danger seems to emanate from below" (Expert Judgement on Markers to Deter Inadvertent Human Intrusion into the Waste Isolation Pilot Plant, Sandia National Laboratories report, SAND92-1382 / UC-721, p. F-49). Credit: Trauth, Kathleen M., Hera, Stephen C., and Guzowsti, Robert V. "Expert Judgement on Markers to Deter Inadvertent Human Intrusion into the Waste Isolation Pilot Plant." Report, Waste Isolation Pilot Program, 1993.

a powerful culture. This place is not a place of honor ... no highly esteemed deed is commemorated here ... nothing valued is here. What is here is dangerous and repulsive to us. This message is a warning about danger.[71]

* * *

Both teams struggled with the long vision required by their mission. Team A noted that, since literacy developed approximately 6,000 years ago and has not ceased to exist, it was reasonable to assume that it will persist into the future. Because of this assumption, there was a major emphasis on written language in the development of markers. Both teams also recognized that political boundaries can change rapidly, so it should not be assumed the United States (if such a thing would still exist in 10,000 years) would hold possession of the territory on which the markers lay. Team B assumed that vandalism might occur and strove to design a marker that would withstand vandalism, or one whose message could still be interpreted if part of the marker were vandalized.

Because of scheduling conflicts Sagan was unable to accept the invitation to join the panel, but he wrote an advisory letter to the committee for publication in the report, along with his suggestions for a marker. In his letter Sagan argued as follows:

> We want a symbol that will be understandable not just to the most educated and scientifically literate members of the population, but to anyone who might come upon this repository. There is one such symbol. It is tried and true. It has been used transculturally for thousands of years, with unmistakable meaning. It is the symbol used on the lintels of cannibal dwellings, the flags of pirates, the insignia of SS divisions and motorcycle gangs, the labels of poisons – the skull and crossbones. Human skeletal anatomy, we can be reasonably sure, will not unrecognizably change in the next few tens of thousands of years.[72]

Yet the skull and crossbones symbol is not as universally interpreted as Sagan suggested. The symbol is generally believed to have roots in early Christian art. In the medieval period, images of skulls and crossbones appeared in paintings of the crucifixion, where they alluded to the apocryphal belief that Christ was crucified atop the buried skull of Adam, the first man. In this tradition, the skull symbolized hope in the form of the imminent resurrection of Christ. In Mexico, Day of the Dead iconography uses a proliferation of skeleton and skull imagery, much of it festive or humorous in tone. The images are not a warning of danger but an invitation to gather and celebrate – not a warning to stay away, but the opposite. In the art history traditions of memento mori and vanitas still life, depictions of skulls are intended to inspire their viewers, inviting them to reflect on their mortality as a reminder that they should make the most of their lives, or should live without vanity or sin. In contemporary emoji use, the skull emoji is generally used by young people to indicate laughing (e.g. "I died from laughter").

We already see that the premise of the universality of this symbol is flawed. But Sagan was so sure of the universal consensus on the skull and crossbones that he concluded his letter thus:

> Unless a more powerful and more direct symbol can be devised, I think the only reason for not using the skull and crossbones is that we believe the cur-

rent political cost of speaking plainly about deadly radioactive waste is worth more than the well-being of future generations.[73]

In this way Sagan highlighted his concern that the DOE might undermine the danger of nuclear waste (and, implicitly, of nuclear weapons) by choosing a message that was not effective in conveying the real danger of this technology, to prevent larger public dissent on nuclear technology. Sagan's concern about the nuclear crisis and his skepticism regarding an effective communication of danger through symbols like the skull and crossbones underscored his pessimism about humanity's ability to address the perils of nuclear technology. The flawed belief in the universality of symbols, as exemplified by the varied interpretations of the skull and crossbones, not only revealed the challenges of communication among humans but also cast doubt on the notion of easily conveying messages to extraterrestrial civilizations. While CETI as a science was imbued with an optimism inspired by fervent internationalism and prolific techno-utopianism, its proponents were not immune to the pessimism brought about by the specter of nuclear annihilation.

The Great Filter

Although I have established that the Cold War in some ways facilitated the development of CETI across the Iron Curtain, Cold War hostilities also led some of its practitioners to lose faith in internationalism. In the earlier part of his career, Shklovsky was hugely optimistic that CETI would show that human beings can overcome their tendencies toward warfare and self-destruction. As an aside at the end of a chapter of *Intelligent Life in the Universe*, Sagan commented:

> At this point in the Russian edition of the present work, Shklovskii expresses his belief that civilizations are not inevitably doomed to self-destruction . . . The political dichotomies of the twentieth century may seem to our remote descendants no more exhaustive of the range of possibilities for the entire future of mankind than do, for us, the alternatives of the European religious wars of the sixteenth and seventeenth centuries. As Shklovskii says, the forces of peace in the world are great. Mankind is not likely to destroy itself. There is too much left to do.

This optimist view of Earth's future mirrored his optimistic position on the likelihood of detecting extraterrestrial intelligence. Toward the end of his life, however, after no success in contacting extraterrestrial intelligence, Shklovsky began publicly expressing doubts about the possibility of CETI's finding evidence of intelligent life in the universe. In one interview, which took place shortly before his death, he stated: "I do not share the opinion of most of my colleagues who believe that the cosmos is totally populated by intelligent beings. I believe life, and especially intelligent life, is an extremely rare phenomenon."[74]

Why the sudden change of heart? It is difficult to say without direct testimony from Shklovsky, of which there is little on the subject. According to his friends, Shklovsky's change of heart was emotional, not scientific. Lev Gindilis wrote that in the later part of his life, Shklovsky "came to a feeling of extreme pessimism in everything, the expression of which was the idea of the loneliness of our civilization, and later – of a dead-end path associated with the acquisition of reason."[75] Shklovsky had become depressed by the state of the world, and according to Gindilis "could not come to terms with what he saw on Earth," and hence "lost a hopeful perspective."[76] Gindilis' account is substantiated by a conversation Drake had with Nikolai Kardashev. According to Kardashev, "Shklovsky had become depressed by the global political situation . . . and concluded that nuclear war was inescapable."[77] He believed that political leaders were too ignorant and foolish to avoid nuclear catastrophe, and it therefore seemed "likely that intelligent aliens had quickly destroyed their worlds, too. Technological civilizations must be a short-lived phenomenon, and we might as well abandon hope of ever finding any."[78]

It is a sobering reading today, in 2023, when the Russian aggression in Ukraine indeed provokes discussion about the use of nuclear weapons. Iosif Shklovsky's concern about the longevity of technological civilizations shows how, when CETI scientists postulated the existence of life in the universe, they began by considering life on Earth. In 1992, seven years after Shklovsky's death, Drake optimistically wrote that, if Shklovsky had lived long enough to witness glasnost and the de-escalation of the arms race, "he would have changed his mind" and "renewed enthusiasm for CETI."[79] Yet in 2024 we again have reason to worry that the Earth's civilizations will destroy themselves. CETI's introspection brings value in the light it sheds on the challenges of our own world, but this very introspection also hinders scientists' ability to think expansively about the variety of civilization in the universe. Their expectations of extraterrestrial intelligence were too intimately tied to their projections of the Earth's future.

I opened this chapter with Enrico Fermi's now famous words, "Where is everybody?" These words led to the development of a theory in C/SETI called the Fermi paradox. The Fermi paradox poses a question regarding the apparent absence of extraterrestrial beings in the vast expanse of the universe. Despite the assumption that life may be widespread, one may well ask why we have not encountered or received any communication from other civilizations. Some have argued that there is a "great filter" in the universe that prevents intelligent civilizations from making contact. Oxford philosopher Nick Bostrom has described the Great Filter as a "probability barrier." The Great Filter, he tells us, "consists of one or more highly improbable evolutionary transitions or steps whose occurrence is required in order for an Earth-like planet to produce an intelligent civilization of a type that would be visible to us with our current observation technology." The Great Filter either lies behind us, meaning that we have overcome some highly improbable event that allowed our civilization to develop, or "it might be ahead of us, somewhere in the millennia or decades to come."[80] Many CETI scientists developed ideas for possible future filters. The first, of course, was the atom bomb. Sagan and Shklovsky were two of the first scientists to raise concerns about the Great Filter when they asked: "Do technical civilizations tend to destroy themselves shortly after they become capable of interstellar radio communications?"[81]

Colonialism was another proposed Great Filter during the Cold War period. An influential paper by the CETI theorist Michael Hart[82] argued that, because humanity "has explored and colonized every portion of the globe it could," it was natural to assume that at least one alien civilization, if it existed, would have colonized the galaxy by 1975.[83] Since there was no evidence that this had happened, Hart argued that humans were likely the first or the only intelligent civilization in the galaxy, which implicitly suggested that we might become the galactic filter ourselves, by eventually colonizing the galaxy. Hart's suggestions highlight the determinism that underscores much of early CETI theory. In making his claim, Hart assumed that, because some cultures on Earth engaged in colonialism, the inevitable unilineal evolutionary track would lead humanity to conquer the cosmos eventually.

Hart's paper was particularly influential in the CETI community. In a 1978 paper exploring the ubiquity of intelligent life in the universe, Shklovsky referenced Hart and agreed that, given that civilizations "must take to the road of unlimited expansion," it may be unlikely that there were many intelligent civilizations near the Earth, since there was no apparent evidence of them.[84] In a chapter titled "Social Evolution" in a NASA History volume, Denning explored contemporary SETI scientists' popular reliance on unilineal evolution and noted that SETI ideas "about extraterrestrial Others are deeply infused with thought about social evolution on Earth."[85] As a result, she argued, scientists "tend to develop syntheses that pull all human experience together into a single narrative."[86] This determinism is ubiquitous in Cold War CETI: fears for our own planet are projected onto extraterrestrial ones.

But this is not simply a cosmic mirror showing us a mirror image of civilization on Earth; it specifically highlights the biases of its practitioners. As we saw in the previous chapter's messages to extraterrestrial intelligence, Cold War CETI was in many ways uniquely Soviet and American in nature. The United States and the Soviet Union were imperial nations; therefore the Soviet and American CETI theorists universalized imperialistic tendencies, and this culminated in our fear of a colonial Great Filter. The United States and the Soviet Union were great technological and spacefaring civilizations at risk of being destroyed by the atom bomb; hence the concern was that all technological and spacefaring civilizations would face the same risk. The cosmic mirror "refracts" instead of reflecting – it's not a really a mirror but a cosmic prism. Just as a prism separates pure light into its component wavelengths, allowing us to determine the identity and structure of the material the light passes through, so too can we use CETI to understand the various elements that make up civilizations on Earth. It allows us to better understand not a universal truth or the "pure light" about Earth's culture, but the different wavelengths of concerns, hopes, fears, challenges, and accomplishments of specific cultures at a specific moment in time. In studying the cosmic prism, we learn more about the medium through which the ideas of CETI scientists passed, affecting their science.

The act of projecting both our hopes and our fears onto extraterrestrials seems a lasting feat in C/SETI, one that changes in nature over time. I would argue that existential fear is still a major driver in contemporary SETI; it is simply that the types of fear change as our societies do. In the summer of 2022 I helped organize the first biannual SETI conference at the Penn State Center for Extraterrestrial Intelligence. No scientist proposed using nuclear weapons as a means of communication – but the nature of our global anxieties has shifted over time. Today, a major cause of existential fear is global climate change, and this fear is reflected in new search proposals. During the conference, a number of papers raised the subject of "pollution SETI," searching exoplanet atmospheres for signs of pollution caused by technology such as an abundance of nitrogen dioxide (NO_2): this is a general byproduct of the combustion process and might be prolific in the atmospheres of civilizations that have major industrial processes like those on Earth.[87]

Today one of the popular candidates for the Great Filter is artificial intelligence (AI). AI is a contemporary example of an existential fear that has affected SETI theory – specifically, "superintelligent general artificial intelligence with destructive goals."[88] A paper published by National Intelligence University researcher Mark M. Bailey in 2023, shortly after the rise of OpenAI technology prompted a global conversation on the benefits and perils of AI, argued that AI might be the cosmic Great Filter:

> This idea considers the possibility that advanced AI will effectively behave as a second intelligent species with whom we will inevitably share this planet. Considering how things went the last time this happened – when modern humans and Neanderthals coexisted – the potential outcomes are grim.[89]

Bailey's paper is an excellent one with which to see the CETI's refractive qualities in action: here is Bailey, a member of the intelligence community, raising concerns about the existential threat posed by a burgeoning technology, one used both for war and for the advancement of our civilization, and projecting a deterministic perspective rooted in a superficial engagement with Earth history. We may not learn much about extraterrestrial civilizations from such papers, but by observing the deterministic projections onto extraterrestrial intelligence we trace CETI's roots in the Cold War through its being a proxy for the intelligence community, through its connection to burgeoning technologies, and through speculation arising from existential fear.

The unique character of Cold War era CETI followed in part from the onset of the existential crises fostered by the cosmic prism. As he was dying from cancer in the 1990s, Carl Sagan wrote *Billions and Billions*, his memoir. Much of the book covers Sagan's existential concerns, including on climate change. In one chapter he writes explicitly about the anxiety that stems from global concerns:

> Everyone experiences at least a dull background level of assorted anxieties. They almost never go away entirely. Most of them are of course about our everyday lives. There is a clear survival value to this buzz of whispered reminders . . . The trick, if you can pull it off, is to pick the right anxieties. Somewhere between the cheerful dolts and nervous worrywarts there's a state of mind we ought to embrace.[90]

The history of CETI was inextricably tied up with the Cold War, presenting many anxieties and posing communication challenges to the scientists involved, but also forcing them to confront larger philosophical questions. In some sense, the search for technosignatures as we know it today could not have existed without the enormous investment of US and Soviet governments into CETI's major infrastructures and institutions, for instance via the establishment of the National Radio Astronomy Observatory, the funding of conferences, or support for the publication of books. Similarly, without the Space Race there would probably have been far less public support for, and individual interest in, pursuing the CETI problem. Furthermore, it was the Cold War mindset that influenced CETI scientists such as Frank Drake to raise the question of the longevity of civilizations; and that led others, such as Sagan and Morrison, into antiwar activism. But although the Cold War played a role in the establishment of CETI, it also presented many challenges to those who pursued CETI and highlighted the cultural and communication barriers between scientists in the United States and the Soviet Union. By striving to overcome interterrestrial cultural and communications difficulties and in recognizing the existential dilemma posed by the Cold War, CETI scientists in some sense engaged far more in philosophical and historical problems than in technical ones. In an interview in 1981, Nikolai Kardashev hinted at this idea:

INTERVIEWER: ...Is there anything else that you would like future historians to know about you?
KARDASHEV: I want to say ... that it is possible to discuss in the frame of the SETI problem, *all* problems connected with the best organization for living on the Earth.
INTERVIEWER: You make better people.
KARDASHEV: Yes.[91]

In Kardashev's view, CETI worked toward solving more problems than just the existence of extraterrestrial intelligence. It forged friendships across national gaps, prompted reflection on inequity and politics, and forced its practitioners to confront some of the most vital existential challenges that face our world. It made better people.

Conclusion

This book opened with an anecdote about a submarine and a crisis of communication. I will now conclude with another.

In 1981, a Swedish fisherman discovered an enormous Soviet submarine, equipped with nuclear weapons, beached on the south coast of Sweden.[1] An international crisis ensued before it was discovered that Sweden was not under attack: rather the Soviet submarine had hit a rock that had scrambled its navigational instruments, forcing it aground. The submarine was escorted by the Swedish Navy back to international waters in the Baltic Sea. Though officially branded an accident, the event prompted much fear and speculation in Sweden, and reports of periscope and submarine sightings became regular occurrences on the Baltic coast. As a result, the Swedish Navy began to use helicopters and boats to lower hydrophones into the water, to try to find the submarines allegedly being spotted. Soon the Navy began to hear the "typical sound" of Soviet subs all over Swedish waters.[2] Whenever a hydrophone picked up evidence of this sound, helicopters would drop charges into the sea, hoping to either provoke the subs to rise to the surface or damage them.

Despite years of pursuing the subs, frequent reports of sightings, and the typical sound being constantly heard, a sub never materialized. By 1994, frustration boiled over and Carl Bildt, the Swedish prime minister, angrily confronted the Russian president, Boris Yeltsin, who denied Russian military presence in Swedish waters. At this point several academics – including two biologists, Magnus Wahlberg and Hakan Westerberg – were invited to listen to the previously classified military intelligence recordings of the alleged sound of Russian submarines, to determine their source. After much investigation, Wahlberg and Westerberg discovered that Swedish waters were indeed occupied by something lurking beneath the surface, but it was not a submarine: it was herring.

Most fish have an organ called a "swim bladder," a small pocket filled with gas, which allows fish to regulate how they float in the water. Herring, in massive schools, release this gas through their anal cavities, making a cacophony of bubbling sounds, which the Swedish military had interpreted as the sounds from a Soviet sub. Scientists believe that, in addition to regulating herring's swim bladder, these bubbles are also used as a form of communication within the school. In other words, Sweden bombed schools of flatulating, chatty fish as part of a massive campaign to search for artificial signals from an alien nation.

This event in the history of foreign relations, while superficially comedic (go ahead, laugh), is nonetheless revealing of a major theme in the history of the Cold War, as well as in the search for extraterrestrial intelligence: the act of searching for and identifying intelligent signals from the "other" – when one does not fully understand how to define the motives, culture, or methods of communication of whom or what they are seeking – will inevitably lead to mixed signals.

The Decline of Cold War CETI and the Rise of SETI

As I have briefly alluded, CETI scientists also had a preoccupation with marine communication: the 1961 Green Bank CETI conference was attended by John Lilly, a dolphin biologist who aimed to discover how to create a two-way dialogue between humans and dolphins. It is worth noting that Lilly's project, which had won funding from NASA because of its implications for CETI, faced a disastrous end.[3] Lilly's laboratory lost funding partly on account of his disregard for animal welfare; Lilly, who had a background in neurology, had injected the dolphins with LSD to study the effect of the drugs on cetacean brains and behavior.[4] There were also later concerns over the sexual nature of the relationship between one scientist and the dolphins on the project.[5] Despite the loss of NASA funds, Lilly continued to try to develop dolphin–human communication, using traditional scientific methods as well as some less traditional ones – for instance attempts at telepathy.[6] Lilly's efforts largely failed, and today the discipline of dolphin biology has shifted to trying to understand the complexities of dolphin communication, rather than trying to teach them English or establish direct dialogue. Lilly's laboratory has since faced much mockery, such as by featuring on a lewd *Saturday Night Live* sketch titled "The Dolphin Who Learned to Speak."[7]

The problems with Lilly's laboratory reflect one of the biggest issues of mid twentieth-century CETI: primarily, the anthropocentric assumption that human conceptions of communication could be universalized across species (or even across different nations). And there is no shortage of unusual research projects such as Lilly's in the history of CETI. For example, perhaps taking inspiration from Lilly's use of LSD and telepathic communication, CETI radio astronomer Gerrit Verschuur once claimed to have taken LSD and enclosed himself in a sensory deprivation tank to try to contact extraterrestrial intelligence with his mind.[8] Perhaps because of its close ties to pseudoscience and a history of some maverick ideas, and also partly as a result of what is sometimes called the "giggle factor," scientific CETI has sometimes struggled with achieving serious recognition within the scientific community and from funding bodies. The giggle factor refers to the skepticism or dismissiveness that some people (including other scientists) exhibit when discussing the search for intelligent alien life; throughout its history, CETI was often met with amusement or ridicule on account of the speculative nature of the field and the absence of concrete evidence.

There are several sources for the giggle factor's impact on CETI. For one, the discipline is sometimes co-opted by unidentified flying object (UFO) enthusiasts, much to the irritation and frustration of C/SETI scientists. Second, C/SETI scientists' evocation of popular themes in science fiction sometimes caused the two to become entangled with each other. And, third, because the premise of the field is inherently non-falsifiable, C/SETI is sometimes discredited as non-scientific. Non-falsifiability means that, while one could definitively prove the existence of extraterrestrial intelligence through evidence such as an artificial extraterrestrial signal, the non-existence of extraterrestrial intelligence cannot be proved; and Karl Popper has contributed greatly to the recognition of non-falsifiability a staple (if problematic) criterion for demarcating non-scientific from scientific knowledge. The giggle factor has contributed to a dearth of funding for C/SETI, especially after the Cold War period, when there was a decline both in interest in the intelligence applications of C/SETI technology and in the desire to fund scientific exchanges between the United States and the Soviet Union.

CONCLUSION

During the Cold War, CETI did receive minor support from the US government. The US government and some of its agencies funded CETI conferences, telescope design proposals, workshops, and even minor observational projects. Yet it is important to note that at the end of the Cold War, in the 1990s, American SETI suffered several setbacks. In the 1980s both planning and development for the first NASA-funded SETI observational project, the High-Resolution Microwave Survey (HRMS), were underway. The project officially launched in 1992, but the following year, in September 1993, Senator Richard Bryan of Nevada introduced an amendment during a congressional meeting on NASA's budget that both defunded HRMS and removed SETI from NASA's mission altogether. That same year, the National Science Foundation (NSF) issued a prohibition against funding searches for extraterrestrial intelligence in its annual NSF Guide to Programs.[9] When introducing the amendment to end SETI funding, Senator Bryan exclaimed, "this hopefully will be the end of Martian hunting season at the taxpayer's expense."[10]

The sudden decline in government-funded C/SETI can be explained in a number of ways. First, Jill Tarter has long noted that SETI is at a disadvantage by comparison with other sciences in that it has to ask for further funds (for more searches) on the basis of what is perceived as a set of failures (the absence of detections).[11] Second, by the 1990s, the Earth was becoming more radio-quiet. Intelligence operations were shifting from signals intelligence to cybersecurity through the rapid development of computers and, in popular culture, video killed the radio star: the rise of cable television meant that all that nonsense on television was no longer reaching the cosmos. Quite possibly there was a recognition, both in the government and in the SETI community, that radio was not necessarily the modus operandi for cosmic communication; and there was certainly less of a need to rely on SETI for conducting signals intelligence. Finally, the Fermi paradox was generally misunderstood by government officials; by the 1990s there was a growing sense that, if a detection had not been made after 30 years of C/SETI, perhaps there were no extraterrestrials to be found. President Clinton's science advisor stated in 1994 that "we've done a lot of observing and listening already; and if there were anything obviously here, I think we would have gotten some signal" by that time.[12] This was based on a misunderstanding of how many C/SETI searches had been conducted since 1960, but nevertheless impacted the defunding of CETI. In the Soviet Union, where CETI generally enjoyed more support from state funds, there was also a drop in CETI activity in the 1990s, perhaps to some extent for similar reasons, but primarily because of the dissolution of the Soviet Union. In a paper that charts C/SETI research in the Russian, Soviet, and post-Soviet space, Gurvits and Gindilis admit that "the general decline of science in Russia after the collapse of the Soviet Union could not but affect the state of SETI" and led to an overall decline in CETI activity after the Cold War.[13]

After C/SETI lost government support in the 1990s, it turned to private support from the megarich – a trend not exclusive to SETI but common in the twenty-first-century space sciences more broadly, as evidenced by the development of private aerospace enterprises such as Elon Musk's Space X and Jeff Bezos' Blue Origin. When attempting to build the first radio telescope array dedicated to SETI, scientists approached Paul Allen, co-founder of Microsoft, to fund what later became the Allen Telescope Array, the first array built exclusively for SETI purposes. More recently, the most ambitious SETI observational project in history launched in 2015, with the goal of surveying 1,000,000 stars over ten years.[14] The project, Breakthrough Listen, was given $100 million in funding from Russian Israeli billionaire philanthropist Yuri Milner. Nevertheless, outside private financing, government support for SETI has overall declined since the twentieth century, although this could change yet again.[a]

The Pen Is Mightier Than the Sword

Yet SETI continues to transform as a field. It is presently experiencing a renaissance of sorts as a result of the rise of exoplanetary science and the search for biosignatures. Biosignatures are measurable indicators – for example, of the presence of certain molecules in an exoplanet atmosphere – that provide evidence of biological processes or of the existence of life. When CETI first began, the first exoplanets had not yet been detected.

---

[a] There is some evidence to this effect: with the rise of exoplanetary telescopes and with SETI's rebranding into technosignature research, governments appear to be open to funding search for extraterrestrial intelligence once again. For example, several next-generation radio telescopes, including the Square Kilometer Array and China's FAST, cite technosignature searches as part of their science goals.

CONCLUSION

Today there are many instruments that can detect exoplanets, and we now know of thousands of existing exoplanets (in fact it is now assumed that every star in our galaxy has at least one planet). Because of this new and exciting development, SETI has been slightly rebranded as a search for technosignatures, playing off as a companion to biosignatures. But there are many in the field who still believe that "SETI" is the best term to use to describe the field. Jason Wright, SETI scientists and the director of the Penn State Center for Extraterrestrial Intelligence, has justified his preference for the term SETI in defining the field:

> To be sure, SETI certainly includes technosignature searches, but is broader than that, much in the way astrobiology is about much more than just searches for biosignatures. "SETI" also properly refers to the goal of the search (finding another "intelligent" species in the cosmos) rather than the means to that goal (the technosignatures that point to that species). The primary drawback of the name is that it contains the term "intelligence," and so brings along a whole host of baggage involving the nature of intelligence, its connection to technology, and cultural expectations about the nature of the beings the field is searching for . . . For now, we will simply state that we will use the term historically, acknowledging the field's foundations, but also metonymically, inviting the listener to consider the nature of beings that could create technology we can detect across the vastness of space.[15]

I agree with Jason in thinking that keeping the name SETI is wise, especially for its metonymic value. Metonymy is a figure of speech in which a word or a phrase is used to represent something else, which is closely related or associated with it, and that use is often based on one shared characteristic or attribute. For example, in the idiom "the pen is mightier than the sword," the word "pen" does not refer to an actual pen but stands in metonymically for the broader concept of the power of the written word. In "SETI," the word "intelligence" allows us to reflect on the deeper implications and complexities associated with the search for extraterrestrial intelligence; those implications and complexities encompass not only the technological aspects represented by technosignatures but also the profound questions about the nature of intelligence and civilization, including (and perhaps especially) our own. I would also suggest that, in CETI, the word "communication" has metonymic value too, since it prompts researchers to think expansively about the nature of communication in a variety of forms, as evidenced by the parallels they created between communication across the Iron Curtain and communication with extraterrestrials.

Fundamentally, CETI was about communication faced with insurmountable barriers – a characteristic that also defined the Cold War. In fact, research on the nature of alien communication and civilization dominated that period even outside the CETI community. The economist Thomas Schelling won the Nobel Prize for work that applied game theory concepts to Cold War conflict. Schelling is perhaps best known for developing the concept of mutually assured destruction – the belief that the most effective deterrent to a devastating war is to make your adversary believe that they would not survive the complete annihilation caused by your retaliatory actions. The wisdom behind mutually assured destruction was simple: the most effective way to prevent or resolve conflicts was to actively seek to comprehend and empathize with the viewpoint of one's adversary. Given that clear communication and empathy with an alien viewpoint were critical to resolving global conflicts, Schelling dedicated a significant amount of time to researching how to communicate in the face of tremendous barriers to communication.

In *The Strategy of Conflict*, Schelling introduced a thought experiment: imagine that you are supposed to meet a stranger in New York City, but you have no means of communication with them.[16] How are you supposed to know where and when to meet? It might look, in theory, like an impossible situation. There are 24 hours in the day and thousands of different locations in the bustling city. The likelihood of two strangers unable to communicate randomly choosing the same place and time must be nearly zero. But when Schelling conducted this mental experiment with his students, he found that the most common answer was "noon at (the information booth at) Grand Central Terminal." How could this be? His theory centered on something he called "focal points," now largely known as "Schelling points," which described the locations people tended to converge on in the absence of communication or explicit agreement, on the basis of their shared expectations and understanding of the situation.

Jason Wright argued in a 2020 paper that Schelling points are an excellent strategy for SETI.[17] According to him, "an essential component of identifying Schelling points is finding points of commonality ... [but] we have very little we can be sure we have in common with an alien species."[18] Wright suggests that a solution to this communication barrier is for SETI searchers to search for "natural" landmarks in space, points in space that are akin to cosmic Grand Central Stations. In fact, as Wright points out, Schelling cited CETI scientists Morrison and Cocconi in *The Strategy of Conflict*, arguing that the 21-centimeter hydrogen line is an example of a cosmic focal point. As Wright and Schelling emphasize, communication often relies on what we share, not on what makes us different. As I explained throughout this book, during the Cold War period there were many barriers to scientific communication. CETI scientists shared a disciplinary grounding and a desire to seek intelligent life in the cosmos, as well as a goal of achieving global harmony. For Soviet and American scientists who tried to cooperate toward these goals, CETI itself became a sort of Schelling point, a field in which they could meet over a shared goal and ideal.

Fake News

As already noted, the Soviet historical record for astronomy and CETI is much more challenging to access than the American one, in part because of the many contemporary barriers, both political and organizational, to archival access in Russia. In my efforts to fill in the gaps of Soviet contributions to radio astronomy, I stumbled upon information on KRT-10, an alleged space radio interferometer launched in 1979. Despite seeing a few references to the existence of the telescope, including its depiction on a Russian stamp, I could not find any significant scientific publications resulting from data from the telescope or any information about the receiver, frequency range, or scientific mission, other than vague references to its alleged ability to conduct very long baseline interferometry (VLBI), a technique in radio astronomy that significantly enhances the resolution of astronomical observations by combining signals from multiple distant antennas.[19] One *Pravda* article titled "Radio Astronomy Steps into Outer Space" highlighted KRT-10 as the first radio interferometer in space. The article further noted that interferometry experts Gennadii Sholomitskii (also of CTA-102 fame) and Leonid Matveyenko were responsible for first proposing VLBI as a method.[20]

I found this article's claims bizarre, primarily because there is another, more contemporary mission, RadioAstron (launched in 2011), that is better known as Russia's first space radio interferometer. Furthermore, I could not find any scientific papers that purported to use VLBI data from KRT-10. RadioAstron, on the other hand, was the product of decades of collaboration between scientists in the United States, especially NASA and NRAO, and Kardashev's research group in the Soviet Union, then in Russia after the fall of the Soviet Union in 1991. RadioAstron was made possible in large part thanks to the ongoing relationships between the United States and the Soviet Union that stem from the first joint VLBI experiments conducted in the late 1960s. Although this idea was not included in promotional or scientific materials, it is the general understanding of scientists at the Astro Space Center located in the Russian Space Research Institute in Moscow, which launched RadioAstron, that RadioAstron was to some extent the culmination of Nikolai Kardashev's dream of using VLBI to seek evidence of extraterrestrial intelligence.[21]

* * *

In the autumn of 2019, I was invited to visit the Astro Space Center and attend the final RadioAstron International Scientific Council meeting. RadioAstron had been decommissioned that summer after astronomers had lost contact with the spacecraft in the aftermath of some mechanical failures. During the meeting I had the opportunity to sit down and conduct an interview with my friend and colleague Leonid Gurvits, an ex-Soviet astronomer and aerospace engineer who began his career in Moscow in the early 1970s and had been a key figure in the RadioAstron mission.[22] At the time of this book's writing, Leonid is an emeritus scientist at the Joint Institute for VLBI ERIC in the Netherlands, as well as a professor in the aerospace department at the Delft University of Technology. Given his expertise in VLBI and his employment at the Soviet Space Research Institute in the 1970s, I assumed that he would know more details about the mysterious KRT-10. When asked, Gurvits responded with his characteristic blend of humor and frankness:

> The first large antenna deployed on Salut-6 station in 1979, KRT-10, was presented to the public as a radio telescope and there were even some publications about radio astronomy. Even more, [there were claims that] VLBI

observations [were made] using KRT and ground base telescopes. And that was complete fake. It was really fake news and much more fake than what is referred to by President Trump. Very different. That was real fake.[23]

When pressed on how he could possibly know this, Gurvits explained that he was a member of the science group tasked with building KRT-10 and, he proclaimed, "we didn't do any VLBI, I swear."[24] Gurvits asserted that doing VLBI would have been impossible, as KRT-10 did not have the correct the equipment needed for it, such as an atomic clock. The true function of KRT-10 was never fully explained to the Space Research Institute team. But, given their involvement in its construction, they could extrapolate its function. They knew where the dish on KRT-10 was pointing; it spent much of its mission flying over the oceans, pointed down toward the Earth. And what is the purpose of a radio telescope pointed toward the ocean? Gurvits explained that it is possible that KRT-10 would have been able to detect the movement of ships or submarines in the ocean, and may well have been used to track the movements of adversarial navies. In other words, its status as a radio telescope was meant to conceal its true nature as an intelligence-gathering tool.

I conclude my book with this brief story on KRT-10 because I believe that it further highlights the connections between radio astronomy and global conflict. Interestingly, Leonid himself is a C/SETI enthusiast and has done a significant amount of work trying to preserve Soviet and Russian contributions to the field, including by publishing one of the few existing papers on Soviet CETI history.[25] Leonid, like many of his colleagues, holds internationalist ideas about his science. When I speak with western radio astronomers who engaged in international collaboration during the Cold War era, they often spend much time trying to persuade me of the apolitical nature of their pursuits. They tell me of the friendships they made in the Soviet Union, how they were driven to cooperate out of a pure scientific desire to understand the universe. The astronomer I quoted in the Introduction, Dave Jauncey, made it clear that he believed his work was "separate from politics."[26] Jauncey and others were not being insincere – they did form lifelong friendships and developed exciting techniques in astronomy that created the foundation of the science conducted today. I believe them when they tell me that they were not motivated by politics.

But in speaking with Gurvits and the other Soviet astronomers I recognized the barriers that had to be overcome to facilitate internationally cooperative experiments and meetings. There was also a recognition that things were different for the Soviet side. Today, historians of the Cold War rightfully push back against the US propaganda of an apolitical science and note the many ways in which the United States weaponized ideology in its science.[27] Yet the historical literature is sometimes too bombastic in its attempts to convince readers to correct their perception that the Soviet Union was a more intellectually repressive society than the United States. Because while it is absolutely true that the United States employed ideology, cultural "diplomacy," psychological warfare, and undue military influence in its scientific–technical pursuits, when assessing the evidence, it is still clear that science in the Soviet Union was far more restrictive and repressive, often to the detriment of its practitioners. It is my view that the important historiographical attempts to reverse the narrative of apolitical US science are sometimes an overcorrection. While politics certainly imbued American science, as has been made apparent in this book, it is also clear that astronomers in the Soviet Union faced unique political barriers that impeded their scientific freedom, as evidenced by the lack of credit Sholomitskii received for the discovery of radio variability. This, as we have seen, is due in some measure to the *explicit* connection between Soviet military and scientific infrastructures, as opposed to the *implicit* connections in American ones.

I come to this conclusion not as an American keen on protecting my country's image as a force for freedom and democracy (I will happily critique my nation's history), but rather from listening to the perspectives of my formerly Soviet friends and colleagues. American astronomers, of course, are and were affected by political forces, but the ties between the military and astronomy were more overt in the Soviet Union than in the United States; this situation often caused US astronomers to be more ignorant of the political forces behind their scientific labor than were their Soviet peers, but it also permitted them a greater degree of autonomy – something that Soviet astronomers lacked. This is a key point to be made: US astronomers were more easily able to hold optimistic and internationalist ideals because of the covert nature of US scientific politics. When I informed Gurvits of his American colleagues' sunny recollections of their collaborative work in the Soviet Union, he responded:

> Yeah, as just normal human beings, we wanted to be open with our colleagues and friends. And as you certainly know, all [the American] people you interviewed... they are my good friends and colleagues for many years... [But] when our foreign colleagues tell us "it was so friendly and lovely!"... Yeah, it was. But what does it cost them, on the other side? They can hardly imagine... It was very, very difficult. It was so difficult you cannot imagine how difficult it was from this side.[28]

In writing this book, I set out to understand the relationship between CETI and Cold War infrastructures and ideologies. My subsequent analysis has shown that this apparently marginal, barely funded field of research is unusually revealing of Cold War communication because it involved a speculative use of resources on themes that are of broad cultural significance, through technologies and techniques that are important strategically both in scientific research and in military communication. I have established the origins of CETI and radio astronomy as distinct products of the Cold War by showing how governments invested in radio astronomy facilities and infrastructure, partly to support a burgeoning interest in signals intelligence. I demonstrated how CETI's ties to military intelligence posed great communications challenges between scientists in the United States and in the Soviet Union, as evidenced by the CTA-102 affair. We examined the relationship between I. S. Shklovsky and Carl Sagan and showed how the established ties between CETI and the Cold War led to existentialist anxieties in the scientists, culminating in a preoccupation with the longevity of civilizations. We explored the early attempts at sending communications to extraterrestrials and saw how they frequently had Earth-based audiences. I showed how the major tension in CETI, being both internationalist and dependent on military–governmental infrastructures, was reflective of one of the great ironies of the Cold War. As historian Vijay Prashad has noted, cooperation and peaceful exchanges were part of the arsenal of the Cold War, as "both sides ... pelted each other with arguments about peace."[29] Scientists were able to cooperate thanks to governmental support of exchanges and conferences, but their cooperation faced many barriers that stemmed from this governmental intervention. This tension led some prominent CETI scientists, and especially Carl Sagan, to engage in antinuclear activism and to reflect deeply on the longevity of human civilization.

This book is perhaps the first sustained study solely dedicated to analyzing Cold War-era CETI history, and certainly one of the first to adopt a transnational approach through the inclusion of Soviet sources. There is therefore much need for future studies in this area. As noted in Chapter 3, acquiring Soviet CETI sources is difficult and has only grown more difficult since 2019; future historians will need to figure out how to expand the historical record, given these challenges. This will not always mean finding access to established archives; I noted, for instance, upon visiting Shklovsky's old office at the Sternberg Astronomical Institute at Moscow State University, that there was a tremendous number of documents and files that had seemingly not been touched in decades; there is almost certainly a wealth of Soviet astronomical history buried in those rooms, if one has the time, funds, and ability to sort through it. This book has argued that the search for extraterrestrial intelligence must be especially cognizant of Earth-based concerns. Historical studies can be a valuable tool in understanding how social, cultural, and political elements impact and shape the science.

Beautiful Dreams and Horrible Nightmares

I have noted in the Introduction how science fiction, especially first-contact stories, often made use of metaphors for Cold War anxieties. It would be regretful, then, to conclude without giving some attention to one of the most well known first-contact science fiction novels written by a CETI scientist. In 1985, Carl Sagan published *Contact*, which told the story of a fictional SETI scientist named Dr. Ellie Arroway whose characterization was based on Jill Tarter.[30] In the novel, Arroway discovers an extraterrestrial intelligence message by using the Arecibo radio telescope in Puerto Rico. The novel, while certainly written in the genre of science fiction, spends most of its pages on the conflict and collaboration that resulted from the international effort to decode the mysterious alien message. In one chapter, Dr. Arroway attempts to convince the president of the United States to permit sharing the classified extraterrestrial signal with radio telescopes all over the world. She argues:

The Earth turns. You need radio telescopes distributed evenly over many longitudes if you don't want gaps. Any one nation observing only from its own territory is going to dip into the message and dip out – maybe even at the most interesting parts. Now this is the same kind of problem that an American interplanetary spacecraft faces. It broadcasts its findings back to Earth when it passes by some planet, but the United States might be facing the other way at the time... I don't think any nation can accomplish this project alone. It will require many nations, spread out in longitude, all the way around the Earth. It will involve every major radio astronomy facility now in place – the big radio telescopes in Australia, China, India, the Soviet Union, the Middle East, and Western Europe.

In moments like these, Arroway serves as a mouthpiece for Sagan and his belief in CETI's role as an international science and in its potential to bring about global unity.

The novel, written approximately a decade before the author's death, seems to serve almost as an alternative history: a collection of events and conversations that Sagan imagined might have happened if CETI's labors came to fruition. Although Arroway is the only character explicitly said by Sagan to have a counterpart in a real CETI scientist (in fact the novel is prefaced by a disclaimer: "This book is a work of fiction ... any resemblance to actual events or locales or persons living or dead is entirely coincidental"), it is difficult not to see parallels between Sagan's colleagues and the characters in the book. One character, Vasily Gregorovich Lunacharsky, is a Jewish Soviet scientist and close friend of Arroway, with good humor despite being often barred from traveling outside the Soviet Union. While introducing Lunacharsky's character in the novel, Arroway mentions how she took him shopping around Berkeley once, on a rare occasion when he was able to leave the Soviet Union, and she describes how he purchased a "Pray for Sex" button from one of the many irreverent hippy establishments. In his autobiography, Frank Drake recounts an almost identical story where I. S. Shklovsky, while visiting Berkeley in the 1970s, purchased a "Pray for Sex" button, joking: "in your country, it is offensive in only one way. In my country, it is offensive in two independent ways."[31] Lunacharsky, clearly a stand-in for Shklovsky, is quoted in the novel as having said the same joke.[32] Sagan was also involved in the production of the screenplay for the film based on the novel, which included the addition of the character Kent Clarke, a blind astronomer working with Arroway, who is clearly modelled after Kent Cullers, the actual blind astronomer who worked with Tarter on the multi-channel spectrum analyzer mentioned in Chapter 2. The events in the book also parallel real CETI events; when Arroway's first SETI search is shutdown, she turns to a private billionaire donor to fund the effort.

I mention *Contact* here not simply because it is an interesting novel, or because it is fun to see the parallels between fictional characters and real historical figures, but because it provides direct evidence that CETI scientists such as Sagan saw CETI as both a great unifier – the book and the screenplay explicitly make the case that contact with extraterrestrial intelligence would foster international collaboration – and a discipline that highlighted human conflict. After Arroway's signal is detected, much of the friction in the book comes from politics and religion, which hinder the progress of the CETI scientists. The outside critique of this frustrating duality comes from the fictional extraterrestrials, who tell Arroway at the end of the film: "You're an interesting species. An interesting mix. You're capable of such beautiful dreams, and such horrible nightmares."[33] Evoking the L of Drake's equation, they warn: "in the long run, the aggressive civilizations destroy themselves, almost always."[34] Sagan's aliens themselves are aspirational – they belong to a galactic community, where different species had cooperated to engineer a supercivilization, somewhat like what Kardashev had envisioned as a Type 3 civilization. They fulfil the CETI dream of global cooperation, built on an expansionist, technocratic culture – a dream that essentially embodies Cold War-era science.

Notes

Notes to Foreword
1. Drake, Frank. *Is Anyone Out There?* New York: Delacorte Press, 1992, pp. 29–30.
2. Struve, Otto. "Astronomers in Turmoil." *Physics Today* 13.9 (September 1960): 18–23.
3. Oliver, Bernard. "The Rationale for a Preferred Frequency Band: The Water Hole." In Morrison, P., Billingham, J., and Wolfe, J., eds., *The Search for Extraterrestrial Intelligence: SETI*. National Aeronautics and Space Administration, Scientific and Technical Information Office, 1977.

Notes to Introduction
1. Blanton, T. S. and Burr, W. "The Submarines of October: US and Soviet Naval Encounters during the Cuban Missile Crisis." *National Security Archive Electronic Briefing Book No. 75*. National Security Archive, October 31, 2002.
2. Kikoy, H. "Vasili Arkhipov: Soviet Hero that Prevented WW3." War History Online, July 4, 2004.
3. Buzan, Barry. "America in Space: The International Relations of *Star Trek* and *Battlestar Galactica*." *Millennium: Journal of International Studies* 39.1 (2010): 175–180.
4. "Darmok." Episode 2, Season 5 of *Star Trek: The Next Generation*, directed by Winrich Kolbe, written by Joe Menosky. Paramount Domestic Television, 30 September 1991.
5. Sontag, Susan. "The Imagination of Disaster." *Commentary Magazine*, October 1965, n.p.
6. Interview with David Jauncey held on October 4, 2019 in Moscow, Russia. Stored in the Niels Bohr Library & Archives, American Institute of Physic, College Park, MD.
7. Drake, Frank. *Is Anyone Out There?* New York: Delacorte Press, 1992, p. 115.

8. Wolfe, Audra. *Freedom's Laboratory: The Cold War Struggle for the Soul of Science*. Baltimore, MD: Johns Hopkins University Press, 2018.
9. Vladimirov, L. "Glavlit: How the Soviet Censor Works." *Index on Censorship* 1.3–4 (1971): 31–43.
10. Gordin, Michael D. "Lysenko Unemployed: Soviet Genetics after the Aftermath." *Isis* 109.1 (2018): 56–78.
11. Graham, Loren. *Science and Philosophy in the Soviet Union*. New York: Knopf, 1972.
12. McCutcheon, R. A. "The 1936–1937 Purge of Soviet Astronomers." *Slavic Review* 50.1 (1991): 100–117.
13. Wolfe, *Freedom's Laboratory*.
14. "The Longest Search: The Story of the Twenty-One Year Portrait of the Soviet Deep Space Data Link and How It Was Helped by the Search for Extraterrestrial Intelligence." National Security Archives (Undated, but produced after 1983; declassified September 2011).
15. "Comrades." Episode 1 of *The Cold War* (documentary), produced by Pat Mitchell and Jeremy Isaacs. CNN, 1998.
16. Ibid.
17. Conversation carried out in July 2019 with Anamaria Berea, research investigator at Blue Marble Space Institute of Science.
18. Tarter, Jill. "What If There's Somebody Else Out There?" CNN, April 20, 2010.
19. Ibid.

Notes to Chapter 1
1. Sagan, Carl, ed. *Communication with Extraterrestrial Intelligence (CETI)*. Cambridge: MIT Press, 1973, p. 316.
2. Given the informal nature of the conversation, there is of course no documentary evidence to support the claim. However, I spoke with Frank at a panel we both sat on in 2019 and he told me this. The conversation at the conference was allegedly overheard by Kenneth I. Kellermann, who also corroborated Drake's claims. For a citation of Kellermann's own claims, see Kellerman, Kenneth I., "Memories of Nikolai Kardashev: From Astrophysics to Aliens." *Earth and Universe* 350.2 (2023). The article is in Russian.
3. Interview with Kenneth I. Kellermann held on August 4, 2020 in Charlottesville, Virginia. Stored in the Niels Bohr Library & Archives, American Institute of Physic, College Park, MD.
4. Drake, Frank. *Is Anyone Out There?* New York: Delacorte Press, 1992, p. 115.
5. Wright, Jason T., Sheikh, Sofia, Almár, Iván, Denning, Kathryn, Dick,

Steven, and Tarter, Jill. "Recommendations from the Ad Hoc Committee on SETI Nomenclature." Cornell University, arXiv:1809.06857v1, September 18, 2018. https://doi.org/10.48550/arXiv.1809.06857.
6. "Writing History for a Popular Audience: A Round Table Discussion." Organization of American Historians, n.d. https://www.oah.org/tah/august-3/writing-history-for-a-popular-audience-a-round-table-discussion.
7. For a very well researched book on the many serendipitous discoveries in radio astronomy, I recommend Kellermann, Kenneth and Bouton, Ellen. *Star Noise: Discovering the Radio Universe*. Cambridge: Cambridge University Press, 2023.
8. Dick, S. J. *Life on Other Worlds: The 20th-Century Extraterrestrial Life Debate*. Cambridge: Cambridge University Press, 1998, p. 201.
9. "Radio to Stars, Marconi's Hope." *New York Times*, January 20, 1919.
10. Ibid.
11. Ibid.
12. "Marconi Testing His Mars Signals," *New York Times*, January 29, 1920.
13. Ibid.
14. "Hello Earth! Hello!" *The Tomahawk*, March 18, 1920, Image 6. Digital scan held in the Library of Congress.
15. For more on Frank Drake, see Dick, Steven J. *The Biological Universe: The Twentieth Century Extraterrestrial Life Debate and the Limits of Science* Cambridge: Cambridge University Press, 1999, pp. 419–431.
16. Drake, *Is Anyone Out There?*, p. 19.
17. Ibid, pp. xi–xii.
18. Baum, L. Frank. *Tik-Tok of Oz*. Chicago, IL: Reilly & Britton, 1914.
19. Thank you to Daniel Mills for sourcing the original Oz quote.
20. List of Summer Students. Archives of the National Radio Astronomy Observatory, Student Programs Series, Summer Student Programs Unit.
21. Drake, *Is Anyone Out There?*, p. 19.
22. Warren, J. "Ancient Atomists on the Plurality of Worlds." *Classical Quarterly* 54.2 (2004): 354–365.
23. Huygens, C. *Cosmotheoros: The Celestial Worlds Discover'd: or, Conjectures Concerning the Inhabitants, Plants and Productions of the Worlds in the Planets*. London: Timothy Childe, 1698. English translation of a treatise originally composed in Latin under the title *Kosmotheoros, sive de terris coelestibus, earumque ornatu, conjecturae*. (*Kosmotheōros* is itself the Latin transliteration of the Greek word κοσμοθέωρος, "the one who consults or surveys the universe or the skies"; this compound did not exist in ancient Greek, it was probably coined by Christiaan Hygens himself.)
24. Crowe, Michael J. *The Extraterrestrial Life Debate, 1750–1900*. Cambridge: Cambridge University Press, 1986.

25. Dick, Steven J. *The Biological Universe: The Twentieth Century Extraterrestrial Life Debate and the Limits of Science.* Cambridge: Cambridge University Press, 1999, p. 418.
26. Morrison, Philip and Giuseppe Cocconi. "Searching for Interstellar Communication." *Nature* 184.4690 (1959): 844–846.
27. Dick, Steven J. *The Biological Universe: The Twentieth Century Extraterrestrial Life Debate and the Limits of Science.* Cambridge: Cambridge University Press, 1999, p. 418.
28. Lovell, Bernard. "Appendix." In his *The Exploration of Outer Space.* Oxford: Oxford University Press, 1962: 82.
29. Dick, *The Biological Universe.*
30. Ibid., 414.
31. Ibid., 415.
32. For sample works, see Kellermann, Kenneth, Bouton, Ellen, and Brandt, Sierra. *Open Skies: The National Radio Astronomy Observatory and Its Impact on US Radio Astronomy.* New York: Springer International Publishing, 2020; Sullivan III, Woodruff. *Cosmic Noise: A History of Early Radio Astronomy.* Cambridge: Cambridge University Press, 2009; Agar, Jon. "Making a Meal of the Big Dish: The Construction of the Jodrell Bank Mark I Radio Telescope as a Stable Edifice, 1946–57." *British Journal for the History of Science* 27.1 (1994): 3–21.
33. Dick, The Biological Universe, p. 5.
34. Crowe, *The Extraterrestrial Life Debate,* p. 524.
35. Dick, *The Biological Universe,* p. 5.
36. Ibid., 2.
37. Basalla, George. *Civilized Life in the Universe: Scientists on Intelligent Extraterrestrials.* Oxford: Oxford University Press, 2006, p. 12.
38. Ibid.
39. "Interview with Frank Drake." In Swift, David W., *SETI Pioneers: Scientists Talk about Their Search for Extraterrestrial Intelligence.* Tucson: University of Arizona Press, 1990, p. 57.
40. "Skeptical Sunday: The Gospel According to SETI." SETI Institute. https://bigpicturescience.org/episodes/Skeptical_Sunday_The_Gospel_According_to_SETI.
41. Ibid. (for both Tarter quotations).
42. Sagan, Carl. *The Varieties of Scientific Experience: A Personal View of the Search for God,* edited by Ann Druyan. New York: Penguin Group, 2006, p. xiv.
43. Newton, Isaac. "'General Scholium' from the Mathematical Principles of Natural Philosophy." In his *The Mathematical Principles of Natural Philosophy,* vol. 2, London: Benjamin Motte, 1729, p. 501.

44. Clark, K. J. "This Most Beautiful System." In his *Religion and the Sciences of Origins*. New York: Palgrave Macmillan, 2014, p. 184.
45. Dick, Steven J. "Civilized Life in the Universe: Scientists on Intelligent Extraterrestrials." *Physics Today* 60.1 (2007), here pp. 57–58. doi: 10.1063/1.2709561.
46. Hayes, Robert (@nuclearsciencelover), TikTok, March 28, 2022.
47. Chung, Frank. "UFO Believers Think Aliens Will Stop Nuclear War." *Australian News*, March 11, 2022; Ningthoujam, Natalia. "Divine Intervention: Did Mystery UFO Wipe Out Russian Tanks?" *Latin Times*, March 11, 2022.
48. Michaud, Michael A. G. *Contact with Alien Civilizations: Our Hopes and Fears about Encountering Extraterrestrials*. New York: Copernicus Books, 2006, p. 204.
49. "Jill Tarter: A Scientist Searching for Alien Life." NPR, July 23, 2012.
50. Drake, *Is Anyone Out There?*, p. 160.
51. Larkin, Brian. "The Politics and Poetics of Infrastructure." *Annual Review of Anthropology* 42 (2013): 327–343.
52. Sagan, Carl and Shklovsky, I. S. *Intelligent Life in the Universe*. San Francisco, LA: Holden-Day, 1966, p. 461.
53. "Report of the Ad Hoc Subcommittee on UFO Studies with Respect to IAA SETI Permanent Committee Business." Unpublished manuscript.
54. Charbonneau, Rebecca. "Imaginative Cosmos: The Impact of Colonial Heritage in Radio Astronomy and the Search for Extraterrestrial Intelligence." *American Indian Culture and Research Journal* 45.1 (2021): 71–94.
55. Coyne, Jerry. "*Scientific American* Finds the Search for Extraterrestrial Intelligence Racist and Colonialist." In his blog Why Evolution Is True, August 12, 2022.
56. Krauss, Lawrence. "The Lunatics Are Running the Asylum." *Critical Mass*, August 19, 2022.
57. Suyu, S. H., et al. "HoLiCOW I. Ho Lenses in COSMOGRAIL's Wellspring: Program Overview." *Monthly Notices of the Royal Astronomical Society* 468.3 (2017): 2590–2600.
58. Drake, *Is Anyone Out There?*, p. 110.
59. McNeill, William H. "Journey from Common Sense: Notes of a Conference on Communication with Extraterrestrial Intelligence, Byurakan, Armenia, September, 1971." *University of Chicago Magazine*, 64.5 (1972): 2–14; from the Wellcome Library Archives, Francis Crick (1916–2004), Box 102.
60. As quoted in Wineburg, Sam. "Historical Thinking and Other Unnatural Acts." *Phi Delta Kappan* 80.7 (1999): 488–499, here p. 497.

Notes to Chapter 2
1. Wolfe, Audra. *Freedom's Laboratory: The Cold War Struggle for the Soul of Science*. Baltimore, MD: Johns Hopkins University Press, 2018.
2. Balser, Dana S., Ghigo, Frank D., and Lockman, Felix J. *But It Was Fun*. Green Bank: Green Bank Observatory, 2016: xi.
3. Forman, Paul. "Behind Quantum Electronics: National Security as Basis for Physical Research in the United States, 1940–1960," *Historical Studies in the Physical Sciences* 18.1 (1987): 149–229, here p. 150.
4. Sullivan III, Woodruff. *Cosmic Noise: A History of Early Radio Astronomy*. Cambridge: Cambridge University Press, 2009, p. 79.
5. Transcript of President Dwight D. Eisenhower's Farewell Address (1961). Our Documents. https://multimedialearningllc.wordpress.com/wp-content/uploads/2009/04/president-eisenhower_s-farewell-address.pdf.
6. As quoted in Sullivan III, Woodruff. *Cosmic Noise: A History of Early Radio Astronomy*. Cambridge: Cambridge University Press, 2009, p. 81.
7. Hey, J. "Notes on G. L. Interference on 27th and 28th February." 1942 (Secret), file AVIA 7/3544 (6486), Public Record Office, Kew, England; as cited in Sullivan III, *Cosmic Noise*, p. 81.
8. Hey, J. "Solar Radiations in the 4–6 Metre Radio Wave-Length Band." *Nature* 157.47–48 (1946).
9. Hodgson, John. "Sir Bernard Lovell (1913–2012)." The John Rylands Collection Blog. https://rylandscollections.com/2012/09/24/sir-bernard-lovell-1913-2012.
10. Smith, F., Davies, R., and Lyne, A. "Bernard Lovell (1913–2012)," *Nature* 488.592 (2012): 592.
11. Lovell, A. "Electron Density in Meteor Trails". *Nature* 160.670–671 (1947): 670–671.
12. Agar, Jon. "The New Price and Place of University Research: Jodrell Bank, NIRNS and the Context of Post-War British Academic Science." *Contemporary British History* 11.1 (1997): 1–30.
13. MacDonald, Alexander. *The Long Space Age: The Economic Origins of Space Exploration from Colonial American to the Cold War*. New Haven, CT: Yale University Press, 2017, p. 6.
14. Sullivan III, *Cosmic Noise*, p. 439.
15. Ryle, Martin, as cited in Sullivan III, *Cosmic Noise*, p. 440.
16. Drake, Frank, *Is Anyone Out There?* (New York: Delacorte Press, 1992), p. 25.
17. Sullivan III, *Cosmic Noise*, p. 441.
18. Kellermann, Kenneth, Bouton, Ellen, and Brandt, Sierra. *Open Skies: The National Radio Astronomy Observatory and Its Impact on US Radio Astronomy*. New York: Springer International Publishing, 2020, vol. 1, p. 21.

19. Wolfe, Audra. Competing with Soviets: Science, Technology, and the State in Cold War America. Baltimore, MD: John Hopkins University Press, 2013.
20. Wolfe, Freedom's Laboratory.
21. Jill Tarter, a prominent SETI scientist, has compiled an online tool that tracks all the known searches for extraterrestrial technology: visit https://technosearch.seti.org/radio-list.
22. "West Virginia Code, Chapter 37A." Green Bank Observatory.
23. "FCC Docket No. 11745." In National Radio Astronomy Observatory Archives, Records of NRAO, Directors Office, Series/Spectrum Management Unit/Box #1.
24. Van Keuren, D. K. "Cold War Science in Black and White: US Intelligence Gathering and Its Scientific Cover at the Naval Research Laboratory, 1948–62." *Social Studies of Science* 31.2 (2001): 207–229.
25. US Marine Corps, MCRP 2-10. Distribution Statement A.1, Signals Intelligence, October 6, 2021, p. 311. https://www.marines.mil/portals/1/Publications/MCWP%202-10.pdf?ver=2018-12-20-092916-950.
26. Aid, M. M. and Cees, W. "Introduction on the Importance of Signals Intelligence in the Cold War." *Intelligence and National Security* 16.1 (2001): 1–26, here p. 6.
27. Ibid., p. 10.
28. Ibid., p. 20.
29. Oberhaus, David. "The Cold War Plan to Build Earth's Largest Telescope." *Supercluster*, January 13, 2020.
30. Interview with Kenneth I. Kellermann on August 4, 2020 in Charlottesville, Virginia. Stored in the Niels Bohr Library & Archives, American Institute of Physic, College Park, MD.
31. Wolfe, Competing with Soviets, p. 1.
32. Ibid.
33. Van Keuren, "Cold War Science in Black and White.
34. Bamford, James. "The Agency That Could Be Big Brother." *New York Times*, December 25, 2005.
35. Drake, Frank. "Quantifying Earth's Electromagnetic Leakage." Guest editorial, SETI League, n.d. http://www.setileague.org/editor/leakage.htm.
36. Ibid.
37. Ibid.
38. Ibid.
39. Ibid.
40. Morrison, Philip and Cocconi, Giuseppe. "Searching for Interstellar Communication." *Nature* 184.4690 (1959): 845.
41. Smith et al., "Bernard Lovell."

42. Ibid.
43. Ibid.
44. For more on the construction of the Mark I telescope and on the conflicting agendas of Cold War science, see Agar, Jon. "Making a Meal of the Big Dish: The Construction of the Jodrell Bank Mark I Radio Telescope as a Stable Edifice, 1946–57." *British Journal for the History of Science* 27.1 (1994): 3–21.
45. Phelan, Dominic. "Sir Bernard Lovell and the Soviets." *Spaceflight* 56 (2014): 336–338, here 336.
46. "Rocket Reported Falling Rapidly: British Astronomer Asserts First." *The New York Times*, 23 November 1957.
47. Ibid.
48. Phelan, "Sir Bernard Lovell and the Soviets," p. 336.
49. Graham-Smith, Sir Francis and Lovell, Sir Bernard. "Diversions of a Radio Telescope." *Notes and Records of the Royal Society* 62 (2008): 197–204, here p. 200.
50. Ibid, 199.
51. Ibid, 199.
52. University of Manchester, John Rylands Archive. Memorandum by Sir Bernard Lovell on the Files Covering Contact with Soviet Scientists and Visits to the Soviet Union. 1963 File.
53. Phelan, Dominic. "Sir Bernard Lovell and the Soviets." *Spaceflight* 56 (2014): 337.
54. University of Manchester, John Rylands Archive. Memorandum by Sir Bernard Lovell.
55. Kardashev, N. S. and Marochnik, L. S. "The Shklovsky Phenomenon." *Astronomical and Astrophysical Transactions* 30.2 (2017): 119–124, here 119.
56. Shklovsky, I. S. "Emission of Radio-Waves by the Galaxy and the Sun." *Nature* 159 (1947): 752–753.
57. Kardashev et al., "The Shklovsky Phenomenon," p. 120.
58. Friedman, Herbert. "Introduction." In Shklovsky, I. S., *Five Billion Vodka Bottles to the Moon*. New York: W. W. Norton, 1991, p. 29.
59. Sagan, Carl and Shklovsky, I. S. *Intelligent Life in the Universe* (San Francisco, CA: Holden-Day, 1966), p. 362.
60. University of Manchester, John Rylands Archive. Memorandum by Sir Bernard Lovell.
61. Ibid.
62. Ibid.
63. Ibid.
64. Ibid.

65. Hodgson, "Sir Bernard Lovell."
66. "Radiation: Radar." World Health Organization, November 2, 2007. https://www.who.int/news-room/questions-and-answers/item/radiation-radar#:~:text=Environmental%20RF%20levels%20from%20radars,the%20earliest%20known%20health%20effects.
67. Vizzini, Bryan E. "Cold War Fears, Cold War Passions: Conservatives and Liberals Square off in 1950s Science Fiction." *Quarterly Review of Film and Video* 26.1 (2008): 28–39.
68. Frankenheimer, John, dir. *The Manchurian Candidate*. Santa Monica: M. C. Productions, 1962.
69. Hoover, J. Edgar. *Masters of Deceit: The Story of Communism in America and how to Fight It*. New York: Holt, 1958, p. 93.
70. Wolfe, *Freedom's Laboratory*.
71. "MK-Ultra." *History*, June 16, 2017.
72. Office of the Director of National Intelligence. "Updated Assessment of Anomalous Health Incidents." March 2023.
73. Ibid.
74. "Radio List." The SETI Institute. https://technosearch.seti.org/radio-list.
75. Project Cyclops (1971). Papers of Kenneth I. Kellermann, National Radio Astronomy Observatory Archives, National Radio Astronomy Observatory, Charlottesville, VA.
76. Interview with Jill Tarter conducted by Rebecca Charbonneau on September 4, 2019 in Berkeley, CA. Stored in the Niels Bohr Library & Archives, American Institute of Physic, College Park, MD.
77. Ibid.
78. Billingham, John. "SETI: The NASA Years." In Vakoch, Douglas, ed., *Archaeology, Anthropology, And Interstellar Communication*. Washington, DC: NASA History Office, 2014, p. 9.
79. Interview with Jill Tarter conducted by Rebecca Charbonneau.
80. Ibid.
81. Ibid.
82. Aid and Cees, "Introduction on the Importance of Signals Intelligence in the Cold War," p. 1.
83. "The Longest Search: The Story of the Twenty-One Year Portrait of The Soviet Deep Space Data Link and How It Was Helped by the Search for Extraterrestrial Intelligence." National Security Archives (undated, but produced after 1983; declassified September 2011).
84. Ibid., p. 3.
85. Interview with Jill Tarter conducted by Rebecca Charbonneau.
86. "The Longest Search," p. 2.
87. "Stonehouse: First US Collector of [REDACTED] Signals." National

Security Agency (undated but declassified September 2007); "The Longest Search," p. 2.
88. Ibid.
89. Ibid., p. 3.
90. Ibid.
91. Interview with Jill Tarter conducted by Rebecca Charbonneau.

Notes to Chapter 3
1. Denning, Kathryn. "Impossible Predictions of the Unprecedented: Analogy, History, and the Work of Prognostication." In Vakoch, D., ed., *Astrobiology, History, and Society*. New York: Springer, 2013, p. 310.
2. Charbonneau, Rebecca. "Imaginative Cosmos: The Impact of Colonial Heritage in Radio Astronomy and the Search for Extraterrestrial Intelligence." *American Indian Culture and Research Journal* 45.1 (2021): 71–94.
3. Denning, "Impossible Predictions of the Unprecedented," p. 310.
4. Ibid., p. 302.
5. Kellermann, Kenneth I. Preface to Braude, S. Y. et al., eds., *A Brief History of Radio Astronomy in the USSR*. New York: Springer, 2012: 40.
6. Ibid., p. vi.
7. Ibid., p. v.
8. Gingerich, Owen. "The Central Bureau for Astronomical Telegrams." *Physics Today* 21.12 (1968): 36–40, here p. 37.
9. Ibid., p. 37.
10. Ibid., p. 40.
11. Ibid., p. 39.
12. Kellermann, Kenneth, and Pauliny-Toth, I. I. K. "Variable Radio Sources." *Annual Review of Astronomy and Astrophysics* 6 (1968): 417–448, here p. 417.
13. Sholomitskii, G. B. "Variability of the Radio Source CTA-102." *Information Bulletin on Variable Stars* 83 (1965).
14. Kellermann, Preface, p. vii.
15. Sholomitskii, "Variability of the Radio Source CTA-102."
16. TASS telegram written by Alexander Midler, April 12, 1965 (in Russian). Records of the Telegraph Agency of the Soviet Union (TASS), Centre for Preservation of a Reserve Record, Ialutorovsk, Siberia, Russia. Scans courtesy of Leonid Gurvits.
17. Kellermann and Pauliny-Toth, "Variable Radio Sources," p. 421.
18. Ibid.
19. Ibid., p. 420.
20. Kellermann and Pauliny-Toth, "Variable Radio Sources," p. 421.

21. Kellermann, Preface, p. viii.
22. Thronson, H. A., Hawarden, T. G., Davies, J. K., Penny, A. J., Orlowska, A., Sholomitskii, G., Saraceno, P., Vigroux, L., and Thompson, D. "The Edison International Space Observatory and the Future of Infrared Space Astronomy." *Advances in Space Research* 18.11 (1996): 171–183.
23. University of Manchester, John Rylands Archive. Memorandum by Sir Bernard Lovell on the Files Covering Contact with Soviet Scientists and Visits to the Soviet Union. 1963 File.
24. Kellermann, Kenneth I. "The Discovery of Quasars and Its Aftermath." *Journal of Astronomical History and Heritage* 17.3 (2014): 267–282.
25. Heidmann, Jean. "SETI False Alerts as 'Laboratory' Tests for an International Protocol Formulation." *Acta Astronautica* 21.2 (1990): 73–80, here p. 74.
26. Kellermann, "The Discovery of Quasars and Its Aftermath," p. 267.
27. Ibid., p. 267.
28. Roth, A. "Putin Approves Law Targeting Journalists as 'Foreign Agents.'" *The Guardian* (Moscow), December 3, 2019.
29. Gordin, Michael. "The Forgetting and Rediscovery of Soviet Machine Translation." *Critical Inquiry* 46 (2020): 835–866, here p. 837.
30. This section on Russian cosmism draws on research conducted during my Master of Science degree at Oxford: Charbonneau, R. "Examining Intelligent Life in the Universe: How SETI Internationalism Facilitated Scientific Collaboration during the Cold War." MA Thesis, University of Oxford, 2017.
31. Young, George. *The Russian Cosmists: The Esoteric Futurism of Nikolai Fedorov and His Followers*. Oxford: Oxford University Press, 2012, p. 151.
32. Ibid., p. 3.
33. Ibid., p. 4.
34. Ibid.
35. Hagemeister, Michael. "Russian Cosmism in the 1920s and Today." In Rosenthal, Bernice Glatzer, ed., *The Occult in Russian and Soviet Culture*. Ithaca, NY: Cornell University Press, 1997, p. 185.
36. Young, *The Russian Cosmists*, p. 204.
37. Pryakhin, V. "Russian Cosmism and Modernity" (in Russian). *Herald of the Russian Academy of Sciences* 82.6 (2012), p. 474.
38. Bobrovnikoff, Nicholas. "Soviet Attitudes Concerning the Existence of Life in Space." In Wukelic, George E., ed., *Handbook of Soviet Space-Science Research*. New York: Gordon and Breach Science Publishers, 1968, here p. 456.
39. Engels, Friedrich. *Dialectics of Nature*, translated by C. Dutt. Moscow: Progress Publishers, 1954, p. 25.

40. Gindilis, L. M. and Gurvits, L. I. "SETI in Russia, USSR, and the Post-Soviet Space: A Century of Research." *Acta Astronautica* 162 (2019): 1–13.
41. Ibid., p. 6.
42. Kardashev, N. S. "The Communication of Information by Civilizations on Other Worlds" (in Russian). *Astronomicheskii Zhurnal* 41 (1964). Some translate the word Передача in the title as "transmission," others as "broadcast." Передача can be interpreted in several ways, depending on context; I believe either interpretation to be correct but, given that the subject of this book is communication, I opted for the first.
43. Ibid., 218.
44. Wikipedia, s.v. "Kardashev Scale." Last updated on May 13, 2021.
45. Kardashev, "The Communication of Information by Civilizations on Other Worlds," p. 219.
46. Ibid.
47. Interview with Freeman Dyson. In Swift, David W., *SETI Pioneers: Scientists Talk about Their Search for Extraterrestrial Intelligence*. Tucson: University of Arizona Press, 1990, p. 325.
48. Kardashev, "The Communication of Information by Civilizations on Other Worlds," p. 219.
49. Ibid., 220.
50. Ibid.
51. Ibid.
52. Ibid.
53. Gindilis and Gurvits, "SETI in Russia."
54. Graham, Loren. *Science in Russia and the Soviet Union*. Cambridge: Cambridge University Press, 1994, p. 100.
55. Ibid.
56. Sagan, Carl and Shklovsky, I. S. *Intelligent Life in the Universe*. San Francisco, CA: Holden-Day, 1966, p. viii.
57. Tovmasyan, G. M., ed. *Extraterrestrial Civilizations (Vnezemnye tsivilizatsii): Proceedings of the First All-Union Conference on Extraterrestrial Civilizations and Interstellar Communication*. Jerusalem: Israel Program for Scientific Translations, 1967, p. 97.
58. Ibid., 97.
59. Ibid., 30.
60. Ibid., 32.
61. Ibid., 32.
62. Interview with Lev Gindilis on October 3, 2019, in Moscow. Stored in the Niels Bohr Library & Archives, American Institute of Physic, College Park, MD.
63. TASS telegram written by Alexander Midler, April 12, 1965. See n. 16.

64. Ibid.
65. Ibid.
66. Ibid.
67. Ibid.
68. Ibid.
69. Ibid.
70. Ibid.
71. Shklovsky, I. S. *Five Billion Vodka Bottles to the Moon*. New York: W. W. Norton, 1991, p. 253.
72. Ibid., p. 253.
73. "Russians Temper Report on Space." *New York Times*, April 14, 1965.
74. "Interview with Nikolai Kardashev." In Swift, David W., ed., *SETI Pioneers: Scientists Talk about Their Search for Extraterrestrial Intelligence*. Tucson: University of Arizona Press, 1990, p. 195.
75. Drake, Frank. *Is Anyone Out There?* New York: Delacorte Press, 1992, p. 103.
76. Choldin, Marianna Tax. "Access to Foreign Publications in Soviet Libraries." *Libraries & Culture* 26.1 (1991): 135–150.
77. The Byrds, "C.T.A. 102" (1967).
78. "Space Claim Is 'Sad.'" *Coventry Evening Telegraph*, April 13, 1965.
79. Sullivan, Walter. "Natural Origins Indicated." *New York Times*, April 13, 1965.
80. Sullivan, Walter. "Radio Emissions from Space Spur Disagreement between Soviet and American Astronomers." *New York Times*, April 18, 1965.
81. Ibid.
82. Ibid.
83. Ibid.
84. Ibid.
85. Kellermann, Preface, p. viii.
86. Ibid.
87. Midler, Alexander. "! OR ?," in Gurvits, L. I., ed., *I. S. Shklovsky: Mind, Life, Universe* (in Russian). Moscow: 2019, p. 215.
88. Gindilis and Gurvits, "SETI in Russia."
89. Interview with Leonid Gurvits on October 4, 2019 in Moscow. Stored in the Niels Bohr Library & Archives, American Institute of Physic, College Park, MD.
90. Scheglov, P. V. "Obituary: Shklovsky, Iosif." *Quarterly Journal of the Royal Astronomical Society* 27.4 (1986): 701.
91. "New in Radio Astronomy" (in Russian). *Pravda*, April 14, 1965.
92. Ibid.
93. Midler, "! OR ?"

Notes to Chapter 4

1. Wolfe, Audra. *Freedom's Laboratory: The Cold War Struggle for the Soul of Science*. Baltimore, MD: Johns Hopkins University Press, 2018, p. 2.
2. Sher, Gerson. *From Pugwash to Putin: A Critical History of US–Soviet Scientific Cooperation*. Bloomington: Indiana University Press, 2019, p. 12.
3. Ibid., 14.
4. Wolfe, *Freedom's Laboratory*, p. 15.
5. Interview with Malcolm Longair conducted by Rebecca Charbonneau on November 4, 2019 in Cambridge. Stored in the Niels Bohr Library & Archives, American Institute of Physic, College Park, MD.
6. Sagan, Carl. *The Cosmic Connection: An Extraterrestrial Perspective*. South Shore, London: Anchor Press, 1973, p. 96.
7. Ibid.
8. Ibid.
9. Ibid.
10. Ibid.
11. Ibid.
12. Ibid., p. 97.
13. Ibid.
14. Ibid.
15. Ibid., p. 98.
16. Ibid., p. 99.
17. Jansky, K. G. "Radio Waves from Outside the Solar System." *Nature* 132.3323 (1933): 1–72, here p. 66.
18. Sullivan III, *Cosmic Noise*.
19. Dick, Steven J. *The Biological Universe: The Twentieth Century Extraterrestrial Life Debate and the Limits of Science*. Cambridge: Cambridge University Press, 1999, p. 418.
20. Van de Hulst, H. "Origin of the Radio Waves from Space." *Nederlandsch Tijdschrift voor Natuurkunde* 11.210 (1945), n.p.
21. "Hydrogen Line," lyrics © 1995 by Dr H. Paul Shuch ("Dr. SETI").
22. Sullivan III, *Cosmic Noise*, p. 397.
23. Shklovsky, I. S. "Monochromatic Radio Emission from the Galaxy and the Possibility of Its Observation" (in Russian). *Astronomicheskii Zhurnal* 26 (1949).
24. Shklovsky, "Monochromatic Radio Emission from the Galaxy," p. 10.
25. Morrison, Philip and Cocconi, Giuseppe. "Searching for Interstellar Communication." *Nature* 184.4690 (1959): 844–846.
26. Shklovsky, I. S. *Five Billion Vodka Bottles to the Moon*. New York: W. W. Norton, 1991, p. 250.

27. Shklovsky, I. S. "Is It Possible to Communicate with Intelligent Beings of Other Planets?" (in Russian). *Priroda* 7 (1960): 21–30.
28. Ibid., p. 21.
29. Ibid., p. 29.
30. "Luna 1." NASA Space Science Data Coordinated Archive. Last uptated October 28, 2022 (version 5.1.15). https://nssdc.gsfc.nasa.gov/nmc/spacecraft/displayTrajectory.action?id=1959-012A.
31. Grahn, S. "Jodrell Bank's Role in Early Space Tracking Activities: Part 1." Jodrell Bank Centre for Astrophysics. University of Manchester, September 16, 2008. https://www.jb.man.ac.uk/history/tracking/part1.html.
32. Shklovsky, *Five Billion Vodka Bottles to the Moon*, p. 248.
33. Shklovsky, I. S., Esipov, V. F., Kurt, V. G., Moroz, V. I., and Shcheglov, P. V. "An Artificial Comet." *Soviet Astronomy* 36.6 (1959): 986–991, here p. 986.
34. Kardashev, N. S. and Marochnik, L. S. "The Shklovsky Phenomenon." *Astronomical and Astrophysical Transactions* 30. 2 (2017): 119–124.
35. Shklovsky, *Five Billion Vodka Bottles to the Moon*, p. 248.
36. Ibid.
37. Ibid., p. 247.
38. Interview with Leonid Gurvits on October 4, 2019 conducted in Moscow. Stored in the Niels Bohr Library & Archives, American Institute of Physic, College Park, MD.
39. Vladimirov, L. "Glavlit: How the Soviet Censor Works." *Index on Censorship* 1.3–4 (1971), p. 31.
40. Ibid., p. 35.
41. Shklovsky, *Five Billion Vodka Bottles to the Moon*, p. 250.
42. Gordin, M. D. "Lysenko Unemployed: Soviet Genetics after the Aftermath." *Isis* 109.1 (2018): 56–78, here p. 73.
43. Shklovsky, *Five Billion Vodka Bottles to the Moon*, p. 250.
44. Ibid., p. 250.
45. Ibid., p. 251.
46. Shklovsky, I. S. Universe, Life, Mind (in Russian). Nuclear Physics on the Internet, 1963. Visit https://www.researchgate.net/profile/Guillermo-Lemarchand/publication/316596339_Intelligent_Life_in_the_Universe_an_example_of_the_scientific_cooperative_endeavor_during_the_sixties/links/5906181aa6fdccd580d37c2d/Intelligent-Life-in-the-Universe-an-example-of-the-scientific-cooperative-endeavor-during-the-sixties.pdf.
47. Stanislav, Lem. *Summa technologiae* (in Russian). Moscow: Mir Publishers, 1968, p. 608.
48. "We Chose to Go to the Moon." John F. Kennedy Moon Speech – Rice

Stadium, September 12, 1962. Johnson Space Center. https://www.facebook.com/watch/?v=10155264732864030.
49. Kohonen, I. "The Space Race and Soviet Utopian Thinking." *Sociological Review* 57 (2009): 114–131, here p. 115.
50. Buzan, Barry. "America in Space: The International Relations of *Star Trek* and *Battlestar Galactica*." *Millennium: Journal of International Studies* 39.1 (2010): 175–180.
51. Mullen, L. "Carl Sagan (1934–1996)." NASA Science Solar System Exploration. November 26, 2001 https://science.nasa.gov/people/carl-sagan.
52. Morowitz, H. "Life on Venus." *Nature* 215 (1967): 1259–1260.
53. Sagan, C. "Direct Contact among Galactic Civilizations by Relativistic Interstellar Spaceflight." *Planetary and Space Science* 11 (1963): 485–498, here 485.
54. Letter from Carl Sagan to I. S. Shklovsky, June 8, 1962. Library of Congress, Carl Sagan and Ann Druyan Archive.
55. Letter from I. S. Shklovsky to Carl Sagan, n.d. Library of Congress, Carl Sagan and Ann Druyan Archive.
56. Letter from Carl Sagan to I. S. Shklovsky, August 6, 1962. Library of Congress, Carl Sagan and Ann Druyan Archive.
57. Letter from Sagan to Shklovsky, June 8, 1962.
58. Friedman, Herbert. "Introduction." In Shklovsky, *Five Billion Vodka Bottles to the Moon*, p. 9.
59. Ibid., p. 17.
60. Interview with Rustam Dagkesamanski conducted on October 8, 2019 in Pushchino, Russia. Stored in the Niels Bohr Library & Archives, American Institute of Physic, College Park, MD.
61. McCutcheon, R. A. "The 1936–1937 Purge of Soviet Astronomers." *Slavic Review* 50.1 (1991): 100–117, here p. 100.
62. Ibid., p. 108.
63. Fitzpatrick, S. *The Practice of Denunciation in Stalinist Russia*. National Council for Soviet and East European Research, Washington, DC, 1994, p. III.
64. Ibid.
65. Fitzpatrick, *The Practice of Denunciation in Stalinist Russia*, p. iii.
66. Ibid.
67. Shklovsky, I. S. "Non-Fiction Stories" (in Russian). *Eniergiia* 6 (1988), pp. 41–42.
68. Shklovsky, *Five Billion Vodka Bottles to the Moon*, p. 17.
69. Ibid., p. 17.
70. Ibid.

71. The Kennedy Proposal for a Joint Moon Flight. NASA History Office, 1963. https://www.hq.nasa.gov/pao/History/SP-4209/ch2-4.htm.
72. Ibid.
73. University of Manchester, John Rylands Archive. Memorandum by Sir Bernard Lovell on the Files Covering Contact with Soviet Scientists and Visits to the Soviet Union. 1963 File.
74. Ibid.
75. Shklovsky, *Five Billion Vodka Bottles to the Moon*, p. 251.
76. University of Manchester, John Rylands Archive. Memorandum by Sir Bernard Lovell.
77. Ibid.
78. Friedman, "Introduction," p. 16.
79. University of Manchester, John Rylands Archive. Memorandum by Sir Bernard Lovell.
80. Gordin, "Lysenko Unemployed," p. 58.
81. Ibid.
82. Ibid., p. 59.
83. Some of the research for this section was conducted during my MSc degree in the history of science, medicine, and technology at Oxford. To read my master's dissertation, see Rebecca Charbonneau, "Examining *Intelligent Life in the Universe*: How SETI Internationalism Facilitated Scientific Collaboration during the Cold War," MSc thesis, University of Oxford, August 2017.
84. Letter from Frederick H. Murphy to I. S. Shklovsky, January 9, 1963. Library of Congress, Carl Sagan and Ann Druyan Archive.
85. Letter from I. S. Shklovsky to Carl Sagan, September 26, 1963. Library of Congress, Carl Sagan and Ann Druyan Archive.
86. Gilman, B. A History of the Deliberate Interference with the Flow of Mail: The Cases of the Soviet Union and the People's Republic of China: A Report. Washington, DC: US Government Printing Office, 1989, p. vii.
87. Ibid., p. 1.
88. Ibid., p. 2.
89. Pessen, E. "Appraising American Cold War Policy by Its Means of Implementation." *Reviews in American History* 18.4 (1990): 453–465, here p. 453.
90. Ibid., p. 460.
91. United States Congress, House Committee on Post Office and Civil Service, Subcommittee on Postal Operations and Services. Soviet Disruption of Mail: Hearing before the Subcommittee on Postal Operations and Services of the Committee on Post Office and Civil Service, House of

Representatives, One Hundredth Congress, Second Session. Washington, DC: US Government Printing Office, 1988.
92. Gilman, A History of the Deliberate Interference with the Flow of Mail, p. 4.
93. Tabarovsky, I. "Understanding the Real Origin of That New York Times Cartoon." *Tablet*, June 6, 2019. https://www.tabletmag.com/sections/arts-letters/articles/soviet-anti-semitic-cartoons.
94. Ibid.
95. Letter from I. S. Shklovsky to Carl Sagan, September 29, 1962. Library of Congress, Carl Sagan and Ann Druyan Archive, in Russian.
96. Shklovsky, *Five Billion Vodka Bottles to the Moon*, p. 251.
97. Letter from Carl Sagan to I. S. Shklovsky on October 8, 1963. Library of Congress, Carl Sagan and Ann Druyan Archive.
98. Sagan, Carl and Shklovsky, I. S. *Intelligent Life in the Universe*. San Francisco, CA: Holden-Day, 1966, p. vii.
99. Shklovsky, *Five Billion Vodka Bottles to the Moon*, p. 252.
100. Ibid.
101. Sagan and Shklovsky, *Intelligent Life in the Universe*, p. vii.
102. Ibid., p. viii.
103. Shklovsky, *Five Billion Vodka Bottles to the Moon*, p. 252.
104. Swift, David W. SETI Pioneers: Scientists Talk about Their Search for Extraterrestrial Intelligence (Tucson: University of Arizona Press, 1990), p. 214.
105. Ibid., p. 214.
106. "Science: Utter Bilge?" *Time*, January 16, 1956.
107. Library of Congress, Correspondence between I. S. Shklovsky and Carl Sagan. The Seth MacFarlane Collection of the Carl Sagan and Ann Druyan Archive. Box 27, Shklovsky, I. S.
108. Öpik, E. J. "Book Review: *Intelligent Life in the Universe*." *Irish Astronomical Journal* 8 (1967), p. 94.
109. Book Review: *Intelligent Life in the Universe*. *The American Biology Teacher* 30.4 (1968), p. 336.
110. Burr, W., ed. "How Many and Where Were the Nukes? What the US Government No Longer Wants You to Know about Nuclear Weapons During the Cold War." National Security Archive, Electronic Briefing Book No. 197, 2006. https://nsarchive2.gwu.edu/NSAEBB/NSAEBB197/index.htm.
111. According to some estimates, even in a best-case scenario for attack by an aggressor state, more than 100 nuclear detonations would have devastating environmental impacts, which are sometimes collectively labelled "nuclear winter" or "nuclear autumn." See Pearce, J. M. and Denkenberger D. C.

"A National Pragmatic Safety Limit for Nuclear Weapon Quantities." *Safety* 8 (2018), 25. doi: 10.3390/safety4020025.
112. Sagan and Shklovsky, *Intelligent Life in the Universe*, p. 358.
113. Interview with I. S. Shklovskii. In Swift, *SETI Pioneers*, p. 52.
114. Ibid.

Notes to Chapter 5
1. Dumas, Stéphane. "Message to Extra-Terrestrial Intelligence: A Historical Perspective." August 17, 2015. https://www.researchgate.net/publication/281036518_Message_to_Extra-Terrestrial_Intelligence_-_a_historical_perspective.
2. Valentine, Geneviève. "You Never Get a Seventh Chance to Make a First Impression: An Awkward History of Our Space Transmissions." *Lightspeed* 10, March 2011.
3. Reynolds, Matt. "The Almighty Tussle over Whether We Should Talk to Aliens or Not." *Wired*, September 26, 2018. https://www.wired.com/story/messaging-aliens-seti-meti (see also https://chinaheritage.net/journal/you-are-garlic-chives-trisolarans-burn-book-and-chinas-men-in-black).
4. Dumas, "Message to Extra-Terrestrial Intelligence."
5. "In the Words of the Cosmos: Lenin, USSR, Peace." *Krasnaia Zvezda* (*Red Star*), December 30, 1962.
6. Knapp, Alex. "Apollo 11: Facts, Figures, Business." *Forbes*, July 20, 2019.
7. Strodder, Chris. *The Disneyland Encyclopedia*, 3rd edn. Santa Monica, CA: Santa Monica Press, 2017, pp. 477–479.
8. Kennedy, John F. "Address at Rice University on the Nation's Space Effort." John F. Kennedy Presidential Library. For this text, see also n. 49 in chapter 4.
9. Riabchikov, Evgenii. "Volia k pobede" ("The Will to Win"). *Aviatsiia i kosmonavtika* (*Aviation and Astronautics*) 4 (1962): 10–19, here p. 19.
10. Wolfe, Audra. *Freedom's Laboratory: The Cold War Struggle for the Soul of Science*. Baltimore, MD: Johns Hopkins University Press, 2018, p. 56.
11. Gerovitch, Slava. "'New Soviet Man' Inside Machine: Human Engineering, Spacecraft Design, and the Construction of Communism." *Osiris* 22.1 (2007): 135–157, here p. 150.
12. Wolfe, *Freedom's Laboratory*, p. 56.
13. Chaikin, Andrew. "Hard Landings." *Air & Space Magazine/Smithsonian*, July 1997. https://www.smithsonianmag.com/air-space-magazine/hard-landings-23935.
14. Mitchell, Don. "Soviet Spececraft Pennants." Mental Landscape, 2004. http://mentallandscape.com/V_Pennants.htm.

15. Harvey, Brian. *Soviet and Russian Lunar Exploration*. Berlin: Springer Praxis Books, 2007, p. 33.
16. Howell, Elizabeth. "Pioneer 10: Greetings from Earth." *Space.com*, September 18, 2012.
17. Sagan, Carl. *Carl Sagan's Cosmic Connection: An Extraterrestrial Perspective*. Cambridge: Cambridge University Press, 2000, pp. 22–23.
18. Ibid.
19. Wolverton, Mark. *The Depths of Space: The Story of the Pioneer Planetary Probes*. Washington, DC: National Academies Press, 2004, p. 80.
20. Sagan, *Cosmic Connection*, p. 25.
21. Interview with Joseph Davis conducted by Rebecca Charbonneau, January 24, 2023, Charlottesville, VA.
22. Davis, Joe. "Monsters, Maps, Signals and Codes." In Bulatov, Dmitry, ed., *Biomediale: Contemporary Society and Genomic Culture*. Kaliningrad: Yantarny Skaz, 2004. http://kaliningrad-old.ncca.ru/biomediale/index-78.html?blang=eng&author=davis.
23. Ibid.
24. Ibid.
25. Gibbs, W. Wayt. "Art as a Form of Life." *Scientific American*, April 1, 2001.
26. 378 U.S. 184 (1964) at 197 (Stewart, J., concurring).
27. *Roth v. United States*, 354 U.S. 476 (1957).
28. Strub, Whitney. "The Clearly Obscene and the Queerly Obscene: Heteronormativity and Obscenity in Cold War Los Angeles." *American Quarterly* 60.2 (2008): 373–398, here p. 374.
29. Ibid.
30. "Poetica Vaginal": Interview with Joe Davis conducted by Benjamin Walker in *Theory of Everything*. National Public Radio, May 17, 2006.
31. "Interstellar Radio Messages." In *David Darling's Encyclopedia of Science*. 2016. https://www.daviddarling.info/encyclopedia/I/IRM.html.
32. Interview with Joseph Davis conducted by Rebecca Charbonneau, January 24, 2023, Charlottesville, VA.
33. Kristeva, Julia. *Powers of Horror: An Essay on Abjection*. New York: Columbia University Press, 1982, p. 4.
34. Austin, Juliette. "Transcending Abjection Through the Body in Performance Art: A Carolee Schneemann and Marina Abramovic analysis." Thoughts of Art, May 11, 2019. https://thoughtsofcapetownart.wordpress.com/2019/05/11/176.
35. Interview with Joseph Davis conducted by Rebecca Charbonneau.
36. Drake, Frank. *Is Anyone Out There?* New York: Delacorte Press, 1992, p. 33.

37. List of Summer Students, 1960. Archives of the National Radio Astronomy Observatory, Student Programs Series, Summer Student Programs Unit.
38. Sagan, Carl. *Murmurs of Earth: The Voyager Interstellar Record.* New York: Ballantine Books, 1979, p. 59.
39. Drake, *Is Anyone Out There?*, p. 33.
40. For more details on the discovery of cosmic masers and Ellen's role, see Kellermann, Kenneth and Bouton, Ellen. *Star Noise: Discovering the Radio Universe.* Cambridge: Cambridge University Press, 2023, p. 203.
41. Also credited with creating the message were Richard Isaacman, then a Cornell graduate student, Linda May, another graduate student, and James C. G. Walker, a member of the Arecibo staff at the time. Carl Sagan also served as a consultant on the project and was in fact the first recipient and interpreter of the message.
42. Drake, *Is Anyone Out There?*, p. 182.
43. As quoted by Whitehouse, David. "Still Waiting for a Call." BBC News, November 8, 1999. http://news.bbc.co.uk/1/hi/sci/tech/521666.stm.
44. Steele, Bill. "It's the 25th Anniversary of Earth's First (and Only) Attempt to Phone ET." Cornell News, November 12, 1999.
45. "Images on the Golden Record." NASA Jet Propulsion Laboratory, n.d. https://voyager.jpl.nasa.gov/golden-record/whats-on-the-record/images.
46. Sagan, *Murmurs of Earth*, p. 3.
47. Ibid., p. 11.
48. Ibid., p. 24.
49. Laszlo, Ervin and the Dalai Lama. "Manifesto on Planetary Consciousness," n.d. Downloadable from https://www.scribd.com/document/59257600/Manifesto-on-Planetary-Consciousness (on subscription).
50. Asimov, Isaac. "Hello, CTA-21: Is Anyone There?" *New York Times*, November 29, 1964.
51. Ibid.
52. Ibid.
53. Drake, *Is Anyone Out There?*, p. 184.
54. Johnson, Daniel. "Greetings, E.T. (Please Don't Murder Us)." *New York Times*, June 28, 2017.
55. "ET, Stay Home," *The Guardian*, January 17, 2008.
56. E.g. Impey, Chris, Spitzer, Anna, and Stoeger, William. *Encountering Life in the Universe: Ethical Foundations and Social Implications of Astrobiology.* Tucson: University of Arizona Press, 2013; Grinspoon, David, *Earth in Human Hands: Shaping Our Planet's Future.* New York: Grand Central Publishing, 2016.
57. Letter from Martin Ryle to Bernard Lovell, July 29, 1976. Ryle Archive, Churchill College, University of Cambridge.

58. As recorded in an anonymous article titled "Protests Continue Abroad," in the *New York Times* of October 23, 1961.
59. Brumfiel, Geoff. "The Strange Story of Project West Ford: Space Junk Before There Was Space Junk." *Wired*, August 28, 2013.
60. Cooper, Mark. "Starfish Prime: The Nuclear Test That Projected Apollo to the Moon." *Wired*, March 9, 2012.
61. Drake, *Is Anyone Out There?*, p. 184.
62. Letter from Ryle to Lovell, July 29, 1976.

Notes to Chapter 6
1. "Enrico Fermi Dead at 53: Architect of Atomic Bomb." *New York Times*, November 29, 1954.
2. Jones, E. M. "'Where Is Everybody?' An Account of Fermi's Question." NASA 1985, Bibcode 1985STIN...8530988J. doi:10.2172/5746675.
3. Denning, Kathryn. "Impossible Predictions of the Unprecedented: Analogy, History, and the Work of Prognostication." In *NASA/ESA Conference on Adaptive Hardware and Systems*, 2013, p. 310.
4. Snyder, C. R. *The Psychology of Hope*. New York: Free Press, 2003.
5. Vakoch, Douglas A., ed. *Psychology of Space Exploration: Contemporary Research in Historical Perspective*. Washington, DC: NASA History Office, 2012, p. 144.
6. Mitchell, Edgar. "Bio: Edgar Mitchell's Strange Voyage." *People Magazine*, April 8, 1974.
7. Suedfeld, Peter, Wilk, Kasia E., and Cassel, Lindi. "Flying with Strangers: Postmission Reflections of Multinational Space Crews." In Vakoch, *Psychology of Space Exploration*.
8. Burke, Edmund. *A Philosophical Enquiry into the Origin of our Ideas of the Sublime and Beautiful*. London: John C. Nimmo, 1757.
9. Sagan, Carl. *Pale Blue Dot: A Vision of the Human Future in Space*. New York: Ballantine Books, 1994.
10. Ibid.
11. Burke, *A Philosophical Enquiry*. Quoted here from the online selections at https://web.english.upenn.edu/~mgamer/Etexts/burkesublime.html#:~:text=Whatever%20is%20fitted%20in%20any,the%20mind%20is%20capable%20of.
12. Letter from Hugh Odishaw to Otto Struve, July 1961. Search for Extraterrestrial Intelligence (SETI) Series, National Radio Astronomy Observatory Archives, Charlottesville, VA.
13. List of attendees, 1961 SETI Conference. Search for Extraterrestrial Intelligence (SETI) Series, National Radio Astronomy Observatory Archives, Charlottesville, VA.

14. Order of the Dolphin members list, 1961 SETI Conference. Search for Extraterrestrial Intelligence (SETI) Series, National Radio Astronomy Observatory Archives, Charlottesville, VA.
15. Davidson, Keay. "California Cosmology: Carl Sagan: The Berkeley Years." *SF Gate*, August 22, 1999.
16. Letter from Hugh Odishaw to Otto Struve, July 1961. 1961 SETI Conference, Search for Extraterrestrial Intelligence (SETI) Series, National Radio Astronomy Observatory Archives, Charlottesville, VA USA.
17. Ibid.
18. "Drake Equation." SETI Institute, n.d. (last updated July 2021). https://www.seti.org/drake-equation-index.
19. Shostak, S. "The Value of 'L' and the Cosmic Bottleneck." In Dick, S. J. and Lupisella, M., eds., *Cosmos and Culture: Cultural Evolution in a Cosmic Context*. Washington, DC: National Aeronautics and Space Administration Government Printing Office, 2012.
20. Denning, Kathryn. "'L' on Earth." In Vakoch, D. A. and Harrison, A. A., eds., *Civilizations beyond Earth: Extraterrestrial Life and Society*. New York: Berghahn Books, 2011, p. 74.
21. This term has featured earlier in my account and is generally known, but this is the place to explain it with some precision. It refers to a period of improved diplomatic relations and a relaxation of tensions between the United States and the Soviet Union. It began in the late 1960s and peaked in the 1970s.
22. NAS-NRC Archives: Central File: ADM: IR: Exchange Programs: USSR: Symposia: Extraterrestrial Intelligence: [1971].
23. Interview with Lev Gindilis conducted on October 3, 2019, in Moscow. Stored in the Niels Bohr Library & Archives, American Institute of Physic, College Park, MD; NAS-NRC Archives: Central File: ADM: IR: Exchange Programs: USSR: Symposia: Extraterrestrial Intelligence: [1971].
24. Ibid.
25. Ibid.
26. The following section on the 1971 US–USSR CETI conference was developed from research undertaken for my Master of Science dissertation at Oxford, 2016–2017.
27. Sagan, Carl, ed. *Communication with Extraterrestrial Intelligence (CETI)*. Cambridge: MIT Press, 1973, p. 353.
28. Although Dyson was born in the United Kingdom, he spent most of his career working for US universities and was an American citizen. He attended the conference as part of the American delegation.
29. Sagan, *Communication with Extraterrestrial Intelligence*.

30. Orgel, Leslie E. and Crick, Francis. "Directed Panspermia." *Icarus* 19 (1973): 341–346.
31. NAS-NRC Archives: Central File: ADM: IR: Exchange Programs: USSR: Symposia: Extraterrestrial Intelligence: [1971].
32. Graham, Loren. *Science in Russia and the Soviet Union*. Cambridge: Cambridge University Press, 1994, p. 174.
33. "Radio List." SETI Institute, n.d. https://technosearch.seti.org/radio-list.
34. Drake, Frank. *Is Anyone Out There?* New York: Delacorte Press, 1992, p. 96.
35. Peters, Benjamin. *How Not to Network a Nation: The Uneasy History of the Soviet Internet*. Cambridge, MA: MIT Press, 2016.
36. Drake, Frank. "Project Ozma." In Kellermann, K. I. and Seielstad, G. A. *The Search for Extraterrestrial Intelligence: Proceedings of an NRAO Workshop Held at the National Radio Astronomy Observatory*. Green Bank, VA: NRAO, 1985, p. 19.
37. Shklovsky, I. S. *Five Billion Vodka Bottles to the Moon*. New York: W. W. Norton, 1991, p. 258.
38. Sagan, *Communication with Extraterrestrial Intelligence*, p. 3.
39. Ibid., p. 353.
40. Drake, *Is Anyone Out There?* p. 115.
41. Ibid.
42. Sagan, Communication with Extraterrestrial Intelligence, p. 353.
43. Drake, *Is Anyone Out There?* p. 115.
44. Sagan, Communication with Extraterrestrial Intelligence, p. 151.
45. Ibid., p. 398.
46. Ibid.
47. Ibid.
48. Ibid., 401.
49. Sakharov, A. D. *Scientific Works* (in Russian). Moscow: Fizmatlit, 2021, pp. 462–466.
50. Gindilis, L. M. "Andrei Ditriyevich Sakharov and Search for Extraterrestrial Intelligence" (in Russian). *Zemlya i Vselennaya (The Earth and the Universe)* 6 (1990).
51. Drake, *Is Anyone Out There?*, p. 115.
52. "Remembering Minsky." *Edge: Conversations*, January 26, 2016.
53. Ibid.
54. Drake, *Is Anyone Out There?*, p. 115.
55. Ibid., pp. 105–106.
56. Ibid.
57. Ibid.
58. Ibid.

59. Ibid.
60. Sagan, Carl. "Nuclear War and Climatic Catastrophe: Some Policy Implications." *Foreign Affairs* 62.2 (1983): 257–292, here at 292.
61. Lewis, Danny. "Reagan and Gorbachev Agreed to Pause the Cold War in Case of an Alien Invasion." *Smithsonian*, November 25, 2015.
62. Sagan, Carl. "The Common Enemy." *Parade*, February 7, 1988. As reprinted in Sagan, Carl. *Billions and Billions: Thoughts on Life and Death at the Brink of the Millennium*. New York: Random House, 1997, pp. 181–192.
63. Ibid.
64. Ibid.
65. Sagan, Billions and Billions, p. 194.
66. Shuch, H. P. "Prof. Phillip Morrison." SETI League, 2005. http://www.setileague.org/admin/philmorr.htm.
67. Morrison, Philip. *Nothing Is Too Wonderful to be True*. Reading, MA: Addison-Wesley, 1995, p. 201.
68. "War and Peace in the Nuclear Age. Dawn. Interview with Philip Morrison, 1986 [1]." GBH Archives, February 26, 1986.
69. Hess, Harry H., Adkins, John N., Benson, William E. et al. "The Disposal of Radioactive Waste on Land: Report of the Committee on Waste Disposal of the Division of Earth Sciences." Washington, DC: National Academy of Sciences-National Research Council, 1957.
70. Trauth, Kathleen M., Hera, Stephen C., and Guzowski, Robert V. "Expert Judgment on Markers to Deter Inadvertent Human Intrusion into the Waste Isolation Pilot Plant." Report, Waste Isolation Pilot Program, osti.gov. November 1, 1993. https://doi.org/10.2172/10117359.
71. Ibid., F-50.
72. Letter from Carl Sagan to Richard Anderson on August 8, 1990. From Trauth et al., "Expert Judgment," Appendix B.
73. Ibid.
74. Shklovsky Interview (in Russian). YouTube, June 19, 2016.
75. Shklovsky, I. S: *Life, Mind, Universe* (in Russian). Moscow: Russian Academy of Sciences, 2019, pp. 270–274, here p. 273. (this is not Shklovsky's book Universe Life Mind, but rather the retrospective edited collection referenced in Ch 3 and published in 2019 by the academy of sciences in Russia, not USSR)
76. Ibid.
77. Ibid.
78. Ibid.
79. Drake, *Is Anyone Out There?*, p. 211.
80. Bostrom, Nick. "Where Are There? Why I Hope the Search for Extraterrestrial Life Finds Nothing." *MIT Technology Review*, May/June

(2008): 72–77, here p. 00. (it was originally published at that journal, but I accessed it here: https://nickbostrom.com/papers/where-are-they/)
81. Sagan, Carl and Shklovsky, I. S. *Intelligent Life in the Universe*. San Francisco, CA: Holden-Day, 1966, p. 358.
82. Hart, Michael H. "Explanation for the Absence of Extraterrestrials on Earth." *Quarterly Journal of the Royal Astronomical Society* 16 (1975): 128–135.
83. Ibid., p. 132.
84. Shklovsky, I. S. "Mind-Endowed Life in the Universe: Can It Be Unique?" (in Russian). *Social Sciences* 2.9 (1978): 199–215.
85. Denning, Kathryn. "Social Evolution: The State of the Field." In Dick and Lupisella, *Cosmos and Culture*, p. 66.
86. Ibid.
87. "NASA Study: To Find an Extraterrestrial Civilization, Pollution Could Be the Solution." NASA. February 10, 2021.
88. Bostrom, Nick. "Where Are They? Why I Hope the Search for Extraterrestrial Life Finds Nothing." *MIT Technology Review*, May–June (2008): 72–77, here p. 00.(same issue as noted above)
89. Bailey, Mark M. "Could AI be the Great Filter? What Astrobiology Can Teach the Intelligence Community about Anthropogenic Risks." arXiv preprint, arXiv:2106.09351 (2021).
90. Sagan, Billions and Billions, p. 89.
91. "Interview with Nikolai Kardashev." In Swift, David W. *SETI Pioneers: Scientists Talk about Their Search for Extraterrestrial Intelligence*. Tucson: University of Arizona Press, 1990, p. 178.

Notes to Conclusion
1. Leitenberg, Milton. "The Case of the Stranded Sub." *Bulletin of Atomic Scientists* 38. 3 (2015): 10–13.
2. Zubko, Marat and Litovkinunder, Viktor. "Swedish Prime Minister Going to Moscow to Resolve 'Periscope Problem.'" CIA. Unclassified, February 3, 1993. https://documents2.theblackvault.com/documents/cia/ufos/C055 17522.pdf.
3. Riley, Christopher. "The Dolphin Who Loved Me: The NASA-Funding Project That Went Wrong." *The Guardian*, June 8, 2014.
4. Ibid.
5. Mosendz, Polly. "How a Science Experiment Led to Sexual Encounters between a Woman and a Dolphin." *The Atlantic*, June 11, 2014.
6. Riley, "The Dolphin Who Loved Me."
7. "The Dolphin Who Learned to Speak." Episode 5, Season 43 of *Saturday Night Live*. The National Broadcasting Company, November 12, 2017.

8. Verschurr, Gerrit L. *Is Anyone Out There? Personal Adventures in the Search for Extraterrestrial Intelligence.* Self-published, 2015.
9. Tarter, Jill et al. "Three Versions of the Third Law: Technosignatures and Astrobiology." Astrobiology Science Strategy for the Search for Life in the Universe. National Academies of Sciences, Engineering, and Medicine. White Paper, 2018.
10. Wright, Jason. "NASA Should Start Funding SETI again." *Scientific American*, February 7, 2018.
11. Garber, Stephen J. "Searching for Good Science: The Cancellation of NASA's SETI Program." *Journal of the British Interplanetary Society* 52 (1999): 3–12, here p. 9.
12. Ibid.
13. Gindilis, L. M. and Gurvits, L. I. "SETI in Russia, USSR, and the Post-Soviet Space: A Century of Research." *Acta Astronautica* 162 (2019). https://www.sciencedirect.com/science/article/abs/pii/S0094576518318393.
14. "Breakthrough Listen." Breakthrough Initiatives, n.d. https://breakthroughinitiatives.org/initiative/1.
15. Wright, Jason. *SETI: Theory and Practice* (forthcoming). Quotation from a draft kindly provided to me by the author.
16. Schelling, Thomas C. *The Strategy of Conflict.* Cambridge, MA: Harvard University Press, 1960.
17. Wright, Jason. "Planck Frequencies as Schelling Points in SETI." *International Journal of Astrobiology* 19.6 (2020): 515–518.
18. Ibid., p. 516.
19. "Radio Astronomy Steps into Outer Space" (in Russian). *Pravda*, September 2, 1980.
20. Ibid.
21. Interview with Lev Gindilis conducted on October 3, 2019 in Moscow. Stored in the Niels Bohr Library & Archives, American Institute of Physic, College Park, MD.
22. Interview with Leonid Gurvits conducted on October 4, 2019 in Moscow. Stored in the Niels Bohr Library & Archives, American Institute of Physic, College Park, MD.
23. Ibid.
24. Ibid.
25. Gindilis and Gurvits, "SETI in Russia, USSR, and the Post-Soviet Space."
26. Interview with David Jauncey conducted on October 4, 2019 in Moscow. Stored in the Niels Bohr Library & Archives, American Institute of Physic, College Park, MD.
27. For a couple of examples of this literature, see Wolfe, Audra. *Freedom's Laboratory: The Cold War Struggle for the Soul of Science.* Baltimore, MD:

Johns Hopkins University Press, 2018; and Wang, Jessica. *American Science in an Age of Anxiety: Scientists, Anticommunism, and the Cold War.* Chapel Hill: University of North Carolina Press, 1999.
28. Interview with Gurvits conducted on October 4. 2019 in Moscow.
29. Prashad, Vijay. *The Darker Nations: A People's History of the Third World.* New York: New Press, 2007, p. xv.
30. "Jill Tarter Elected to American Academy of Arts and Sciences." SETI Institute, April 27, 2021. https://www.seti.org/press-release/jill-tarter-elected-american-academy-arts-and-sciences.
31. Drake, Frank, *Is Anyone Out There?* New York: Delacorte Press, 1992, p. 96.
32. Sagan, Carl. *Contact.* New York: Simon & Schuster, 1986, p. 111.
33. Zemeckis, Robert, dir. *Contact.* Burbank: Warner Brothers, 1997.
34. Sagan, *Contact,* p. 359.

Index

anti-Semitism, 119, 123, 129
Arecibo message, 150–4, 158, 160
Arecibo telescope, 150–5, 158, 160, 202
Arkhipov, Vasili, 1–2
artificial intelligence, 14, 187
Astronomer's Purge, 118–19
Astro Space Center, 84, 198

Basalla, George, 26–9, 31, 35
Bell Burnell, Jocelyn, 17, 20
Bell Labs, 17, 49
Breakthrough Listen, 25, 194
Brown, Robert Hanbury, 62
Byurakan Astrophysical Observatory, 92, 168–9, 172
Byurakan Joint US-USSR CETI Conference, 13–14, 42, 168–75
Byrds, The, 97

California Institute of Technology (CalTech), 82–3, 91, 94, 97–8, 149–50
Central Bureau for Astronomical Telegrams (CBAT), 79
Central Intelligence Agency (CIA), 57, 69, 74
Churchill, Winston, 49
Cocconi, Giuseppe, 24–5, 61, 67, 111, 112–13, 197
Commission for Research and Exploitation of Cosmic Space, 114
communism, 87–8, 92, 102, 113, 116, 137, 157, 163, 178
Contact (book), 29, 202–4
Contact (film), 13, 43, 131
cosmic mirror, 8, 10–12, 71, 75–6, 87, 109, 141, 144, 146–7, 154, 156–7, 161, 163–4, 178, 186
cosmism, 86–9
Crowe, Michael, 25–7, 31–2
CTA-102, 80–3, 91–2, 94–101, 158, 171, 175, 197, 201
Cuban missile crisis, 1, 167, 174

Darling, David, 28
Davis, Joe, 142–7
Denning, Kathryn, 76–7, 162, 167, 196
Department of Energy, 179
Dick, Steven J., 25–7, 29, 32, 110
Drake, Frank, 6, 14, 20–4, 27, 30, 32, 40, 42, 44, 60–1, 86, 89–91, 97, 111–12, 140, 143, 147–8, 150–5, 158–60, 166–7, 170–6, 180, 188, 203–4
Drake equation, 116, 166–7, 173, 204
Durer, Albrecht, 153
Dyson, Freeman, 90–1, 170
Dyson sphere, 90

INDEX

Eisenhower, Dwight D., 46, 107
Elliot, James, 173–4
Engels, Frederick, 88–9

Fermi, Enrico, 162, 185
Fermi paradox, 185, 193
Franklin, Benjamin, 37–8

Gagarin, Yuri, 100, 135, 138
Gauss, Carl Friedrich, 23, 114
Giggle factor, 192
Gindilis, Lev, 89, 95, 99, 100, 168, 170, 184, 194
Ginzburg, Vitaly, 107
Glavlit, 7
Gorbachev, Mikhail, 177
Green Bank (West Virginia), 22, 45, 50–6, 90, 116, 148, 167, 191
Green Bank Telescope (GBT), 51–2
Gundermann, Ellen, 148–150

Hart, Michael, 185–6
Harvard University, 20, 23, 32, 110, 148–9
Havana syndrome, 69–70
Hey, James Stanley, 46–7
historiography, 16–17
Hoover, J. Edgar, 69
Huygens, Christiaan, 23
hydrogen line, 22, 24, 67, 110–12, 140, 197

Intelligent Life in the Universe (book), 102–3, 105, 122, 125–9, 170, 183
International Astronomical Union (IAU), 79, 80–1, 119, 158, 160, 175

Jansky, Karl, 17, 50, 110
Jet Propulsion Laboratory (JPL), 71–2

Jodrell Bank, 24, 47, 50, 63–4, 67, 72–5, 96, 120–1

Kardashev, Nikolai, 84, 89, 90–7, 100–1, 157–8, 168–9, 170, 172, 184, 189, 198, 204
Kardashev's scale of civilizations, 90–2, 158, 204
Keldysh, Mstislav, 67, 113
Kennedy, John F., 1, 115–6, 135
KGB, 58
Korolev, Sergei, 113
KRT-10, 197–9

Large Hadron Collider, 17
Lavender Scare, 145
Lavoisier, Antoine, 37
Lawrence, Earnest O., 17
Lebedev Physical Institute, 49, 66, 78, 107
Lenin, Vladimir, 87, 133–5, 138
Lilly, John, 166, 191–2
Longair, Malcolm, 107–8
Lovell, Sir Bernard, 24, 47, 62–70, 72, 74–5, 79, 82, 95, 98, 107–8, 112, 120–1, 139, 158, 160
Lovell telescope (Mark I telescope), 24, 63, 112
Lomberg, Jon, 155–6, 180
 see also Voyager Golden Record
Los Alamos, 162
LSD, 69, 191–2
Luna-1, 64, 112–3
Luna-2, 139
Lysenko, Trofim, 114, 120–1, 124
Lysenkoism, 114, 121, 124

mail interference, 122–3
Manhattan Project, 166, 178
manifest destiny, 87–8, 116, 134
Marconi, Guglielmo, 18–9

INDEX

McNamara, Robert, 1
McNeill, William, 40, 77, 173
Mesmer, Franz, 37
Michaud, Michael, 30
Midler, Alexander, 95–6, 99–101
Millstone Hill Steerable Antenna, 143–4
Milner, Yuri, 194
Minsky, Marvin, 14, 175
Mitchell, Edgar, 163
MKUltra, 69
Morse Message, 132–4, 138–9, 142, 153–4, 157
Morrison, Philip, 13, 24–5, 61, 67, 111–13, 166, 169, 170, 172, 178–9, 188, 197
Moscow, 1, 5, 63–7, 89, 96–7, 104, 113, 139, 157, 198, 202

National Aeronautics and Space Administration (NASA), 13, 33–4, 47, 70–1, 103–4, 116, 134, 139, 142, 154–5, 163, 166, 178, 186, 191, 193, 198
Ames Research Center, 13, 71
National Radio Astronomy Observatory (NRAO), 22, 43, 45, 47, 48, 51–5, 57–8, 60–1, 78, 81, 85, 148–50, 159, 166, 198
  85–1 telescope, 22–3, 60
  Very Large Array, 43, 83
National Radio Quiet Zone (NRQZ), 52–4, 56–7, 60
National Security Agency (NSA), 57, 60, 73–4
Naval Research Laboratory (NRL), 49, 59
nuclear weapons, 1–2, 8, 10, 30, 45, 59, 103, 109, 129, 156, 160, 167, 173–4, 176, 179, 183–4, 187, 190

Oparin, Alexander, 88, 114, 120–1
Order of the Dolphin, 166
Owens Valley Radio Observatory, 82, 149

Pariiskii, Yuri, 94, 170, 175–6
Payne-Gaposhkin, Cecilia, 148
perestroika, 122
Pioneer missions, 71, 139–41
Pioneer plaque, 140–7, 149, 152, 157–8
Pravda, 95, 100–1, 124, 159, 197
Project Cyclops, 70–1
Project Diana, 58
Project Ozma, 14, 21–5, 32, 51, 50, 86, 89–91, 112, 143, 147–8, 166, 171
Project West Ford (Project Needles), 159
psychological warfare, 69, 136, 200
Pulkovo Observatory, 49, 78, 94, 175
Putin, Vladimir, 84

quasars, 17, 67, 82, 97–9

radar, 17, 45–7, 49, 58, 62, 65, 68, 73, 75, 90, 105, 110, 120, 132, 138, 143–4, 147, 150, 154
radio frequency interference (RFI), 53–6, 60, 74
RadioAstron, 5, 174, 198
Reagan, Ronald, 176–7
Reber, Grote, 50
Ryle, Sir Martin, 48, 158–60

Sagan, Carl, 6, 14, 28–9, 33, 42–4, 74, 83, 102–5, 108–9, 116–7, 121–9, 131, 139–42, 147–9, 154–7, 164, 166, 168–73, 175–8, 180, 182–3, 185, 188, 201–4
Sakharov, Andrei, 119, 120, 174, 176
Salzman-Sagan, Linda, 140–1, 149, 155
SETI Institute, 13, 28

INDEX

Schelling, Thomas, 196–7
scientific exchange programs, 107–9, 168, 192
Shklovsky, I.S., 2, 9, 14, 42, 49, 66–7, 82, 88–9, 93–4, 96–105, 111–27, 129–30, 168–74, 176, 183–6, 201–3
Sholomitskii, Gennadii, 2, 80–3, 94, 96–9, 197, 200
Shostak, Seth, 13
signals intelligence (SIGINT), 57–8, 61, 73–4
Smithsonian Astrophysical Observatory, 63
Soviet Academy of Sciences
Space Race, 2, 5–6, 64, 72, 83, 101–3, 105, 112, 115, 117–18, 139, 188
Space Sciences Board, 166
Sputnik, 2, 5–6, 50, 62–3, 79, 112–15, 128, 144
Stalin, Joseph, 87, 114, 118, 120–1
Starfish Prime, 159–60
Star Trek, 3–4, 10, 152
Sternberg Astronomical Institute, 49, 66, 89, 95–8, 100, 104, 113, 202
Struve, Otto, 48, 166
submarines, 1, 65, 190–1, 199

Sugar Grove, 52, 56–61, 73
600-foot telescope, 57–61

Tarter, Jill, 10, 12–3, 28, 30, 71–2, 74–5, 164, 193, 202, 204
Telegraph Agency of the Soviet Union (TASS), 95–7, 100
Teller, Edward, 59
Tesla, Nikola, 18–9
Three Body Problem (film), 4

UFOs, 26, 30, 32–4, 162, 192
Universe, Life, Mind (book), 112, 114–17, 120–1, 125
US National Academy of Sciences

van de Hulst, Henrik, 67, 110–1
Voyager missions, 71, 154–7, 164–5, 180
Voyager Golden Record, 155–7, 180

Watson-Watt, Robert, 46
Waste Isolation Pilot Plant (WIPP), 179–81

Yevpatoria array (ADU-1000), 65, 68, 80–1, 98–9, 107, 120, 132–3, 138, 154